Flexible Systems Management

The main objective of this series on Flexible Systems Management is to provide a rich collection of research as well as practice based contributions, from different contexts, that can serve as reference material in this upcoming area. Some of these books will be published in association with 'Global Institute of Flexible Systems Management'. It will help in cross-fertilizing ideas from different perspectives of flexibility so as to consolidate and enrich the paradigm of flexible systems management. The audience for the volumes under this series includes researchers, management students/teachers, and practitioners interested in exploring various facets of flexibility research and practice. The series features five types of books:

- *Post conference volumes* containing peer reviewed high quality research papers around a theme and clustered in sub-themes that can act as good reference material.
- *Contributed thematic volumes* based on invited papers from leading professionals, from academia as well practicing world, containing state of the art on an emerging theme.
- *Research monographs* based on research work making a comprehensive contribution to the body of knowledge.
- *Books based on novel frameworks and methodologies* covering new developments that are well tested and ready for wider application in research as well as practice.
- *Business practices and case based* books documenting flexibility practices, strategies, and systems in real life organizations.

The series covers multiple perspectives of flexible systems management; some leading ones, inter alia, are:

- Holistic management of organizational paradoxes with systemic flexibility: including various connotations such as ambidexterity, adaptability, responsiveness, openness, customization, localization, agility, vitality, sustainability, etc.
- Business agility infused by new information and communication technologies: including volatile and virtual business, developments in information and communication technologies generating IT agility such as cloud computing, social networking, knowledge based systems, search technologies, mobile transactions, business continuity, disaster recovery, etc.
- Managing innovation, strategic change and risk: including strategic change, confluence of continuity and change, strategic flexibility, strategy execution, innovation in products/services, processes, management practices, and strategies, business dynamics, business uncertainty and associated risk, etc.
- Flexibility in various operations for achieving business excellence: including organizational flexibility, financial flexibility, manufacturing flexibility, information systems flexibility, marketing flexibility, operational and supply chain flexibility, technology management flexibility, flexibility in business excellence/maturity models, etc.

Review Process

The series follows multi-rung review process where the proposal for each volume is reviewed (single blind) first by the series editor. The series editor enlists the help of editorial board for further comments based on subject matter expertise. Finally, a double blind peer review is conducted by the publishing editor (comments and responses shared with the series editor).

Ethics Statement for this series can be found in the Springer standard guidelines here https://www.springer.com/us/authors-editors/journal-author/journal-author-helpdesk/before-you-start/before-you-start/1330#c14214

Sanjoy Kumar Paul · Renu Agarwal ·
Ruhul Amin Sarker · Towfique Rahman
Editors

Supply Chain Risk and Disruption Management

Latest Tools, Techniques and Management Approaches

Springer

Editors
Sanjoy Kumar Paul
UTS Business School
University of Technology Sydney
Sydney, NSW, Australia

Renu Agarwal
UTS Business School
University of Technology Sydney
Sydney, NSW, Australia

Ruhul Amin Sarker
School of Engineering and Information Technology
University of New South Wales
Canberra, ACT, Australia

Towfique Rahman
UTS Business School
University of Technology Sydney
Sydney, NSW, Australia

ISSN 2199-8493 ISSN 2199-8507 (electronic)
Flexible Systems Management
ISBN 978-981-99-2628-2 ISBN 978-981-99-2629-9 (eBook)
https://doi.org/10.1007/978-981-99-2629-9

This Springer imprint is published by the registered company Springer Nature Singapore Pte Ltd.
The registered company address is: 152 Beach Road, #21-01/04 Gateway East, Singapore 189721, Singapore

Contents

Editors and Contributors

About the Editors

Sanjoy Kumar Paul is the Program Director of the Master of Strategic Supply Chain Management program and an Associate Professor at the UTS Business School, University of Technology Sydney, Australia. Sanjoy's research interests include supply chain risk management, resilience and sustainability, applied operations research, modeling and simulation, and intelligent decision-making. Sanjoy has published more than 120 articles in top-tier journals and conferences, including the *European Journal of Operational Research, Transportation Research Part E: Logistics and Transportation Review, International Journal of Production Economics, Journal of Business Research, International Journal of Production Research, Computers and Operations Research, Computers & Industrial Engineering, International Journal of Logistics Management, Business Strategy and the Environment, Journal of Cleaner Production,* and *Annals of Operations Research,* among many others. He is also an associate editor, area editor, editorial board member, and active reviewer of several reputed journals. Sanjoy has successfully secured external grants from the Department of Defence, Department of Industry, Science, Energy and Resources, Department of Foreign Affairs and Trade, and Meat and Livestock Australia Limited. Sanjoy has received several awards in his career, including ASOR Rising Star Award from the Australian Society for Operations Research, Excellence in Early Career Research Award from UTS Business School, the Stephen Fester prize for most outstanding thesis from UNSW, and high impact publications awards for publishing articles in top-tier journals. Based on his citation records in 2020 and 2021, he was included in the top 2% of scientists in author databases of standardized citation indicators.

Renu Agarwal is a Professor of Management (Strategic, Operations and Supply Chain) and the Director of a new Supply Chain Management Online Program at the University of Technology Sydney. Previously Renu has had extensive industry experience working for State Rail Authority, and Telstra and its joint venture company REACH. Renu has been instrumental in managing several million dollars worth of research projects funded by DIISR Canberra, MED New Zealand, NSW Health, and Queensland Health working in collaboration with LSE, McKinsey and Stanford University which have all contributed to the broader International World Management Survey. One of the many research projects led to the 2009 landmark study—Management Matters in Australia: Just how productive are we?—which has had an impact on Australian government policy making. Recently, Renu led the design of the Australian Management Capability Survey in collaboration with DIISR and ABS which resulted in the release of the data cube output 8172.0—Management and Organisational Capabilities of Australian Business. Besides, Renu has undertaken research for many corporate organizations, including Slyp, Deloitte Access Economics, NSW Innovation and Productivity Council and the Department of Industry, Australian Industry Group, ResMed, SGEC, Microsoft, ASCI, Hargraves Institute, FIAL, CSIRO, Centre for Workplace Leadership, Melbourne University. Renu provides leadership in the disciplinary fields of service innovation, service value networks, supply chain management, dynamic capability building in VUCA environment, management practices, innovation and productivity. She is the editor of several edited books, has published articles in top-tier journals and conferences, and is also an area editor, guest editor, editorial board member, and active reviewer of many reputed journals.

Ruhul Amin Sarker is a Professor in the School of Engineering and IT (SEIT) at UNSW Canberra located at ADFA. He served as the Director of Faculty PG Research (June 2015 to May 2020) and as the Deputy Head of School (Research) of SEIT (2011–2014). He obtained his Ph.D. from Dalhousie University (former TUNS), Canada and his broad research interests are decision analytics, CI/evolutionary computation, operations research, and applied optimization. He is the lead author of the book 'Optimization Modelling: A Practical Approach'. His name appeared on the recent Stanford University list of top 2% of world's scientists-researchers (research fields: Artificial Intelligence and Operations Research).

Towfique Rahman is pursuing his Ph.D. in operations and supply chain management and working as a casual academic in the Management Discipline Group at the University of Technology Sydney, Sydney, Australia. He has a Bachelor of Science degree from the Islamic University of Technology (IUT-OIC) and a Master of Engineering degree from Bangladesh University of Engineering and Technology (BUET). Previously, he worked as a supply chain executive for well-known companies such as MJL Bangladesh limited, a strategic alliance partner of ExxonMobil. Towfique has an impressive publication record, with numerous articles published in esteemed international journals, including the Journal of Cleaner Production, Computers & Industrial Engineering, Energy, Production Planning & Control, Annals of Operations Research, and many more. He works as a reviewer for renowned international journals, such as the Journal of Cleaner Production and the International Journal of Production Research. Towfique's scholarly contributions have been well-received and have garnered significant citations. His research interests focus on resilient and sustainable supply chain management, risk management, energy-efficient supply chain, logistics, industry 4.0, circular economy, simulation, and supply chain network optimization.

Contributors

Renu Agarwal UTS Business School, University of Technology Sydney, Sydney, Australia

Dindayal Agrawal SOIL School of Business Design, Gurugram, Haryana, India

Eman Almehdawe Faculty of Business Administration, University of Regina, Regina, SK, Canada

Amir Hossein Ansaripoor Curtin University, Perth, Australia

Derek Baker UNE Centre for Agribusiness, University of New England, Armidale, Australia;
Food Agility CRC Ltd., Sydney, NSW, Australia

Noraida Azura Darom Department of Mechanical and Manufacturing Engineering, Faculty of Engineering and Built Environment, Universiti Kebangsaan Malaysia, Bangi, Malaysia

Arun Kumar Deshmukh Institute of Management Studies, Banaras Hindu University, Varanasi, India

Ashish Dwivedi Jindal Global Business School, Jindal Global University, Sonipat, O.P, India

Daryl Essam School of Engineering and Information Technology, University of New South Wales, Canberra, Australia

Derek Friday Waikato Management School, The University of Waikato, Hamilton, New Zealand

Marjia Haque School of Engineering and Information Technology, University of New South Wales, Canberra, Australia

Hawa Hishamuddin Department of Mechanical and Manufacturing Engineering, Faculty of Engineering and Built Environment, Universiti Kebangsaan Malaysia, Bangi, Malaysia

Golam Kabir Industrial Systems Engineering, Faculty of Engineering and Applied Science, University of Regina, Regina, SK, Canada

Sneha Kumari Symbioisis School of Economics, Symbiosis International (Deemed University), Pune, India

Steven Alexander Melnyk Department of Supply Chain Management, Broad College of Business, Michigan State University, East Lansing, USA

Stuart Mounter UNE Centre for Agribusiness, University of New England, Armidale, Australia;
Food Agility CRC Ltd., Sydney, NSW, Australia

Tribhuvan Nath RGSC, Institute of Management Studies, Banaras Hindu University, Varanasi, India

Richard Oloruntoba Curtin University, Perth, Australia

Özden Özcan-Top Graduate School of Informatics, Middle East Technical University, Ankara, Türkiye

Sanjoy Kumar Paul UTS Business School, University of Technology Sydney, Sydney, Australia

Damon Proulx Newcastle Business School, College of Human and Social Futures, The University of Newcastle, Callaghan, Australia

Towfique Rahman UTS Business School, University of Technology Sydney, Sydney, Australia

Michael Rosemann Faculty of Business & Law, School of Management, Queensland University of Technology, Brisbane, Australia;
Food Agility CRC Ltd., Sydney, NSW, Australia

Hemendra Nath Roy Industrial Systems Engineering, Faculty of Engineering and Applied Science, University of Regina, Regina, SK, Canada

Suzanne Ryan Newcastle Business School, College of Human and Social Futures, The University of Newcastle, Callaghan, Australia

Hajar SadeghZadeh University of Melbourne, Melbourne, Australia

Ruhul Sarker School of Engineering and Information Technology, University of New South Wales, Canberra, Australia

Yangyan Shi School of Economics, Jiangsu University of Technology, Changzhou, China;
Macquarie Business School, Macquarie University, Sydney, Australia

Arpit Singh Jindal Global Business School, Jindal Global University, Sonipat, O.P, India

Rajesh Kumar Singh Management Development Institute Gurgaon, Gurgaon, India

Michael Billy Sung Consumer Research Lab, Curtin University, Perth, Australia;
Food Agility CRC Ltd., Sydney, NSW, Australia

Firouzeh Rosa Taghikhah Business School, University of Sydney, Sydney, Australia;
Food Agility CRC Ltd., Sydney, NSW, Australia

Prateek Kumar Tripathi Institute of Management Studies, Banaras Hindu University, Varanasi, India

V. G. Venkatesh EM Normandie Business School, Le Havre, France

Alexey Voinov University of Twente, Enschede, Netherlands;
Food Agility CRC Ltd., Sydney, NSW, Australia

Moe Thander Wynn Faculty of Science and Engineering, Queensland University of Technology, Brisbane, Australia;
Food Agility CRC Ltd., Sydney, NSW, Australia

Overview of Supply Chain Risk and Disruption Management Tools, Techniques, and Approaches

Towfique Rahman, Sanjoy Kumar Paul, Renu Agarwal, and Ruhul Sarker

1 Introduction

The influence of supply chain (SC) risk and disruption management on business performance and operations has been widely studied and discussed by scholars and practitioners for the last two decades. In recent years, the focus of SC risk and disruption management studies has shifted from relying on historical data to understanding the probability of occurrence and magnitude for each event that can materially disrupt operations, including identifying the top SC disruptions such as poor supplier performance, forecasting errors, transportation failures, etc. (Paul et al. 2021a, b). These traditional methods do not manage disruptions caused by unforeseen events, as evidenced by the recent Coronavirus (COVID-19) pandemic that caused large-scale and long-lasting disruption; its effects highlight the need for resilient approaches to managing risk and disruption in SCs due to unforeseen events (Chowdhury et al. 2021).

Today's disruptions of SCs are referred to as 'unknown unknown risks', because they are unidentified. The timing and locations of this category of risks are unpredictable despite the fact that businesses are operating in a VUCA (volatile, uncertain, complex, and adaptable) environment. Unfortunately, it is not possible to eliminate

T. Rahman (✉) · S. K. Paul · R. Agarwal
UTS Business School, University of Technology Sydney, Sydney, Australia
e-mail: Towfique.Rahman@uts.edu.au

S. K. Paul
e-mail: Sanjoy.Paul@uts.edu.au

R. Agarwal
e-mail: Renu.Agarwal@uts.edu.au

R. Sarker
School of Engineering and Information Technology, University of New South Wales, Canberra, Australia
e-mail: r.sarker@adfa.edu.au

S. K. Paul et al. (eds.), *Supply Chain Risk and Disruption Management*, Flexible Systems Management, https://doi.org/10.1007/978-981-99-2629-9_1

all risks from unknown unknowns, large and small, in SCs (Modgil et al. 2021). During the recent COVID-19 outbreak, SC operations were adversely affected by an unknown unknown risk on a large scale (Paul et al. 2021a, b). Because of social distancing and lockdowns during the COVID-19 outbreak, manufacturing capacity has been dramatically reduced (Rahman et al. 2021a, b), the distribution and transportation networks have been disrupted severely (Mu et al. 2021), the supply of essential products has been notably disrupted (Gholami-Zanjani et al. 2021), social and environmental sustainability practices have been adversely affected (Nicola et al. 2020), employee's health and safety issues have increased (Rehman and Ali 2021), and the financial performances of SCs have been significantly reduced (Razavian et al. 2021).

The primary focus of SC management is on maximizing profits and returns on investment; hence, managers devote less time and resources to risk and disruption management since they think risk management's return on investment is ignored. The COVID-19 outbreak has proven them wrong. According to a study conducted by the Australasian Supply Chain Institute, only 45% of organizations have developed a business continuity and resilience plan, and only 6.12% of organizations are able to manage risks, rated as 'excellent' in this category (Majumdar et al. 2021). As a matter of fact, most organizations do not have a risk management plan and do not consider the resilience of their SCs. Most managers design their SCs so that cost-efficiency is maximized at the expense of resilience, sustainability, and other risk management practices (Modgil et al. 2021). In spite of the fact that a cost-effective SC may be seen as a lucrative option in the short term, it might not last if managers only focus on maximizing profits and saving money (Gurtu and Johny 2021). Considering the recent COVID-19 pandemic, SC risk and disruption were significant, unpredictable, and disastrous. Therefore, it is important to examine this category of unknown unknown risks as well as the disruptions this category of risks may cause. In order to effectively manage SC risk and disruption, we need greater insight into different dimensions of risks, their causes and sources, tools for assessing them, strategies for ensuring SC resiliency, methods for long-term recovery, management approaches for implementing risk and disruption management tools and techniques, and strategies for overcoming barriers to implementation. Future megatrends require SC business models to change in order to manage risks and disruptions of this category in innovative ways.

This chapter explores SC risks, disruptions, and their effects. This chapter also explores the sources of SC risks, vulnerabilities, and uncertainties. Given that the COVID-19 pandemic has disrupted global SCs and raised numerous problems, this chapter also investigates large-scale SC disruptions and their long-term consequences. The chapter further discusses resilience strategies to manage SC risks and disruption as part of an SC risk management approach. In this section, SC sustainability, adaptability, and viability are also briefly discussed. Finally, tools, techniques, and approaches applied in SC risk management are described in this chapter, so academics and professionals can better understand how to manage SC risks. Conclusions are drawn that highlight future research directions for the study of SC risk management.

2 Supply Chain Risks and Disruptions

Supply chain risks are divided into two categories: "micro risks" and "macro risks" (Gupta and Ivanov 2020). Micro risks in SCs arise primarily as a consequence of daily operational issues such as unexpected supplier failures, equipment failures, manufacturing shut-downs, port congestion, lead time changes, and late delivery owing to transportation limitations (Shekarian and Mellat Parast 2021). Macro risks, on the other hand, are often brought about by large-scale disruptions like natural disasters, disease outbreaks, and pandemics (Can Saglam et al. 2020). Unpredictable accidents in SC networks can destabilize the balance and profitability of SCs; they must be handled to re-establish balance and manage possible risks and disruptions. Unpredictable events might include unexpectedly large orders, demand spikes, supplier supply delays owing to operational risks, production unit malfunctions, and so on (Blackhurst et al. 2018). Risks arise as a result of SC insecurity. Micro and macro interruptions in SC networks may wreak havoc on operations. Natural disasters, geopolitical unrest, terrorist attacks, epidemics, and pandemics all add to the SC's insecurity. Uncertainty creates risks, resulting in the disruption of SCs. Deviations from the planned structure of SC networks are caused by a variety of circumstances. Purposeful and non-purposeful deviations can both impact SC choices, causing SC uncertainty (Macdonald et al. 2018). Theft, terrorism, and financial diversion are examples of purposeful deviations. A non-purposeful divergence may be caused by environmental, economic, or technical factors (Scala and Lindsay 2021). Natural disasters, diseases, and pandemics, such as the recent COVID-19 pandemic, are all examples of unintentional environmental deviations. Non-purposeful economic deviations can have a variety of repercussions, including bull-whip effects and supply-demand variations.

Numerous causes of uncertainty in SCs have been found by researchers, resulting in risks, disturbances, and disruptions. An overview of the causes of SC risk is shown in Fig. 1, taken from the literature (adapted from Ivanov and Sokolov (2010)).

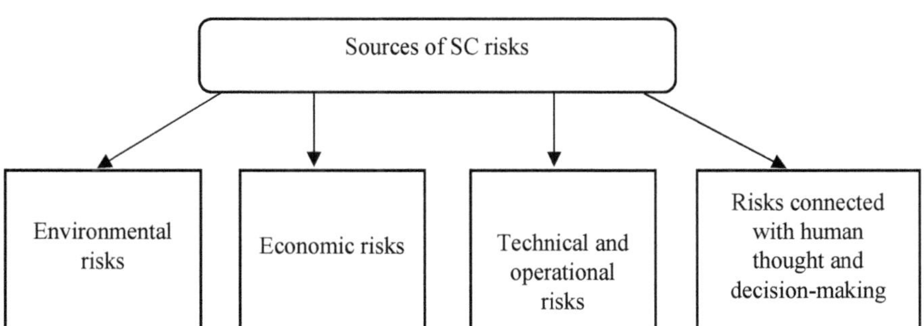

Fig. 1 Sources of risks in SCs. Adapted from Ivanov and Sokolov (2010)

The sources of different SC risks are classified into three groups: (i) internal organizational risks, (ii) SC internal risks, and (iii) external risks (Prakash et al. 2017). The sources of risks in SCs highlighted in Fig. 1 include both internal and external risks.

Environmental risks refer to external, uncontrollable changes. It is difficult to assess and comprehend the extent to which these changes create instability in the usual corporate environment (Christopher et al. 2011). A business SC network needs to understand how a changing environment affects the network, what tactics it can employ to manage the change, and how to maintain a balanced network during the change. Environmental risks include natural disasters, behavioral uncertainties, and goal uncertainty (Majumdar et al. 2021). Uncertainty about the dependability of suppliers, changes in customer preferences and behavior, the unpredictable activities of rivals, changes in product quality, volatility in inter-firm interactions, and other factors can all contribute to SC environmental risks (Kamalahmadi and Parast 2017). Changes in product supply and demand, changes in customer choices and preferences, and changes in technology characterize dynamic environments. The SC network should never overlook the consequences of environmental changes. The demand of consumers, supply from suppliers, technological infrastructure, and rivals within SC networks all have a significant impact on environmental uncertainties. As a result of these environmental uncertainties, demand, supply, manufacturing processes, and SC network control are all unpredictable (Diabat et al. 2013).

Economic uncertainty is a major source of risk throughout SC networks. Economic concerns in SCs include increases in inflation, global recessions, and domestic losses (Cavalcante et al. 2019). The trade conflict between the United States and China, Brexit, worldwide lockdowns, and the COVID-19 pandemic have all had a negative impact on global SCs (Pournader et al. 2020). Businesses cannot control every event outside their doors. In response to external changes, businesses should operate with SCs that are flexible, dynamic, and strategic.

Businesses face various technical and operational uncertainties in their SCs. Examples of technical and operational uncertainties include production failures due to technical insufficiency and problems, lack of experience, etc. (Pavlov et al. 2019). Some operational and technical uncertainty may result in a shortage of capacity, resulting in manufacturers being unable to meet surges in consumer demand. Lacking capacity increases SC's costs. Manufacturers must invest more in high-tech capabilities to respond rapidly to technical and operational uncertainty (Shekarian and Mellat Parast 2021).

Uncertainty in human thought and decision-making can also contribute to SC risks. Such uncertainty is evidenced by inadequate coordination, poor control of logistics, poor decision-making ability, insufficient top-management knowledge, and late top-management decisions (Ivanov and Dolgui 2021b). Human knowledge is extremely valuable in this age of artificial intelligence. Artificial intelligence may produce problems in SC networks that are not guided by human intelligence (Modgil et al. 2021). Managing SC uncertainty necessitates the use of human intelligence and improved decision-making abilities.

Uncertainty causes SC risks, variations, and interruptions. All of these risks cause major disruptions at the macro and micro levels of SCs (supply, demand, manufacturing, transportation, and financial levels). Practitioners must understand the sources of SC risks and uncertainty in order to prevent interruptions. The next section examines the uncertainties, risks, and interruptions that may result from large-scale SC disturbances.

3 Large-Scale Supply Chain Risks and Disruptions

Large-scale risks and disruptions, such as the COVID-19 pandemic, have had a considerable influence on global SC networks, although the degree of their impact is yet unknown. Nonetheless, getting a detailed understanding of the repercussions of SC disruption is critical for developing risk management methods to mitigate the effects. The COVID-19 pandemic is an example of SC disruption on a large scale (Moosavi and Hosseini 2021).

During the global pandemic, SCs worldwide saw significant changes in demand for both high- and low-demand items (Paul and Chowdhury 2020b). As a result, suppliers struggled to ship raw materials to manufacturers in other countries, preventing businesses from ramping up production capacity to match increases in consumer demand. Due to consumers' hoarding behavior and sudden panic-purchasing of high-demand items, supermarkets were badly depleted of necessary supplies (Hall et al. 2020).

Several governments throughout the world established stringent border restrictions, imposed lockdowns, and shut down activities within their borders in an effort to flatten the COVID-19 infection curve (Wang and Yao 2021). Manufacturers struggled to get raw materials from suppliers in restricted (quarantined) zones. Manufacturers typically have only one supplier from a single location. Manufacturing facilities were impacted by supply interruptions and were unable to boost production in order to fulfill consumer demand (Belhadi et al. 2021). As a result, supply interruptions had a significant influence on the whole SC network.

Because the government of most of the countries enforced a restrictive social distancing regulation, most manufacturers were unable to improve their infrastructure to allow their staff to continue working. Due to supply and demand problems, businesses were unable to increase manufacturing capacity. With huge losses and debt, several industries were forced to shut down their manufacturing operations (Choi 2020).

Timely delivery of purchased items to consumers is critical for company SCs to relieve a backlog of orders and related expenditures (Valipour Parkouhi and Safaei Ghadikolaei 2017). Another strategy for firms to preserve goodwill is to deliver items to customers on schedule. Businesses associated with high-demand luxury products had difficulty maintaining quick deliveries due to a shortage of products caused by a manufacturer's low production capacity and lockdowns caused by COVID-19 infections (Rahman et al. 2021a, b).

There was an upsurge in demand for vital supplies as a result of the COVID-19 pandemic inducing fear of lockdowns. Global SCs of critical items failed to foresee the exact need of consumers due to a lack of dynamic demand forecasting skills, technology, and infrastructure. Decision-makers were unable to make timely decisions to restore SC networks due to a lack of knowledge about the exceptional disruption created by the crisis (Hobbs 2020).

Throughout the pandemic, COVID-19 severely disrupted the global SCs of manufacturers. In order to meet consumers' needs, manufacturers were unable to ramp up production. In response, manufacturers of essential products faced increased costs due to shortages. The decline in demand for luxury products led many companies to limit production, reducing profits. Major disruptions in global SCs seriously impacted their financial management (Razavian et al. 2021).

As a result of the COVID-19 pandemic, global SCs have suffered serious degradation in sustainability performance (Rahman et al. 2021a, b). Due to high consumer demand, manufacturers of personal protective equipment had to expand production capacity, such as facemasks, hand-sanitizer, etc. (Rahman et al. 2021a, b). Global SCs suffered significant shortage costs during the closures and shutdowns. As a result of the pandemic-induced global economic recession, many manufacturers had to shut their doors permanently which led to many employees losing their jobs. Companies' reputations were also severely damaged (Karmaker et al. 2021).

4 Managing Supply Chain Risks and Disruptions

Supply chain risk and disruption management is the process of identifying and managing SC risks through a coordinated strategy among SC stakeholders in order to decrease overall SC network vulnerability and disruption (Kilubi 2016). Many scholars have proposed that SC participants employ risk and disruption management process methods to cope with risks and uncertainties posed by or affecting logistics-related activities or resources (Manuj et al. 2014). To maintain profitability and continuity, SC risks are managed by coordination or collaboration among SC participants. Some studies have focused on identifying and managing hazards within the SC network and outside through a coordinated strategy among SC stakeholders to mitigate overall SC vulnerability (Lintukangas et al. 2016). Furthermore, the SC risk management method is distinguished by a cross-company focus on identifying and reducing risks, not only at the business level but also across the whole SC (Wildgoose 2016). In a nutshell, SC risk and disruption management can be defined as "an interorganizational collaborative effort that uses quantitative and qualitative risk management approaches to discover, assess, mitigate, and monitor unanticipated macro and micro level occurrences or situations that might have a negative impact on any portion of a supply chain" (Ho et al. 2015). To manage risk and disruption in SCs, many resilient and sustainable strategies are discussed in the literature. Many dynamic adaptation strategies have recently been suggested to make SCs sustainable and viable in the case of macro-level disruptions, such as the pandemic caused by COVID-19.

5 Resilience Strategies for Managing Supply Chain Risks and Disruptions

The goal of resilience strategies is to make SCs more resilient to disturbance. An efficient SC keeps costs down. A resilient SC, on the other hand, may not be cost-effective at first but it will save firms from interruptions in the long term (Dubey et al. 2019). Researchers are discussing a robust and resilient-sustainable SC. Reconfigurable techniques can aid in the resilience of a healthy and viable SC (DuHadway et al. 2019). During large-scale disruptions, many tiers of SCs may experience simultaneous disturbances, such as supply disruptions, demand disruptions, and logistical disruptions. For example, the COVID-19 pandemic impacted all levels of the SC (Dohale et al. 2021). Due to supply and demand disruptions, producers of essential items such as personal protective equipment (especially facemasks), food, and so on expanded production to fulfill customer demand (Rahman et al. 2021a, b). As a result, there was a significant increase in the waste of critical items such as facemasks in many nations throughout the world. In order to make SCs more viable, resilience strategies must be implemented to maintain social, environmental, and economic performances (Scala and Lindsay 2021). Table 1 describes numerous techniques for increasing the resilience of SCs at supply-level, demand-level, production-level, inventory-level, delivery and transport-level, and financial-level.

6 Supply Chain Sustainability, Adaptability, and Viability

Several scholars have concentrated on resilience methods, sustainability strategies, and so on as means of mitigating the effects of large-scale SC disruptions. Academicians and practitioners have been paying attention to recovery planning for large-scale disruptions (Razavian et al. 2021). The majority of researchers have also focused on response and preparedness strategies (Rahman et al. 2021a, b). Furthermore, most resilience strategies have focused on managing short-term and long-term disruptions. Few strategies were designed to reconfigure SCs from a sustainability point of view (Shishodia et al. 2021). Reconfigurable strategies are those that can readily rearrange SCs in order to survive in a disturbed condition. Despite its importance, sustainable reconfigurable strategies have been little discussed in the literature. The COVID-19 pandemic has demonstrated that global SCs must be redesigned in order to sustain any future extraordinary disruptions so that sustainability and resilience are both maintained (Herold et al. 2021). Decision-makers must employ reconfigurable

Table 1 Selected resilience strategies to manage SC risks and disruptions

SC risks	Resilience strategies	Risk management application to make current SC more resilient	References
Demand risks	Capacity expansion	When there is a rise in demand for vital commodities, an increase in manufacturing capacity may assist in fulfilling that need	Rahman et al. (2021a, b), Ivanov (2020), Ivanov (2019), Ivanov (2021a), Ivanov (2021b), Luthra et al. (2011)
	Purchasing in an emergency	Increasing emergency sourcing will assist in expanding output in order to meet the rise in demand	
	Create new production capacity	Alternative items can be created as a result of the repurposing of manufacturing to suit the temporal need. To fulfill increased healthcare demand during the COVID-19 epidemic, automotive companies, for example, provide valves for respirators	
	Collaborating horizontally and vertically	Horizontal and vertical collaboration can readily promote resource sharing to fulfill consumer needs, particularly during pandemics such as COVID-19	
Supply risks	Alternative sourcing	Alternative sourcing helps maintain the supply in the event of a main supplier breakdown	Dolgui et al. (2018), Dolgui and Ivanov (2021), Ivanov (2021b), Ivanov (2021a), Chowdhury et al. (2021)
	Diverse sourcing	The presence of a diverse set of suppliers improves supply flexibility	
	A local source of supply	Local sourcing allows businesses to be more flexible while also saving money on transportation, which might lead to strong redundancy in the case of a worldwide major disruption such as the COVID-19 pandemic	
	Grouping suppliers by type	When suppliers are categorized by type, it is easier to identify key providers and build emergency preparations	

(continued)

Table 1 (continued)

SC risks	Resilience strategies	Risk management application to make current SC more resilient	References
	Inventory at risk/ Strategic stock	Strategic stock/risk inventory may be beneficial in fulfilling changing customer needs and avoiding stock-outs	
Production risks	Back shoring/ reshoring	We can minimize susceptibility and boost robustness by reshoring and back shoring, which is critical in cases such as the COVID-19 pandemic	Dolgui and Ivanov (2020), Chowdhury et al. (2021), Ivanov (2021b), Paul and Chowdhury (2020a, b), Tarafdar and Qrunfleh (2017), Pavlov et al. (2019), Manuj et al. (2014)
	Local production and nearshoring	Nearshoring and domestic manufacturing help to minimize production vulnerability and boost resilience in the event of a disruption	
	Capacity repurposing	By altering the production system and supply base, repurposing can help to launch rapid demand-supply reallocation	
	Diversifying and substituting products	It may be beneficial to develop a big quantity of alternate items in the event of an SC interruption	
	The postponement of products	Manufacturers can respond swiftly to fluctuating client demand and increase inventory efficiency by deferring production	
	Developing decentralized manufacturing systems	When a major disruption occurs, such as the COVID-19 pandemic, dispersed manufacturing facilities improve resilience	
	Flexible and modular product lines	This aids in responding to varying customer demands during interruptions	
	Facility for subcontracting	Subcontracting permits production to continue in the case of an interruption at the principal manufacturing plant	
	Make use of idle capacity	Allows emergency items to be manufactured	

(continued)

Table 1 (continued)

SC risks	Resilience strategies	Risk management application to make current SC more resilient	References
	3D printing	Produces a variety of products and services in order to keep supply and demand in check	
	The industrial revolution 4.0	Robotic-enabled smart manufacturing facilities can use digital twins to regulate production and scheduling based on real-time data	
	Robotics and human collaboration	In the face of extreme disruptions, such as the COVID-19 pandemic, the use of robotics in production may boost capacity	
Transportation and delivery risks	Collaborate with other transporters	Collaboration with other transportation providers helps to improve the robustness of product delivery to retailers and consumers when a major disruption occurs	Gunasekaran et al. (2015), Paul et al. (2017), Ivanov (2021a), Aldrighetti et al. (2021), Chowdhury et al. (2021), Ishfaq et al. (2021)
	Increase the number of distribution centers	Distributing closer to consumer zones strengthens logistics and enables seamless delivery during disasters	
	Multiple-mode and multi-route shipments	Multimodal and multi-route shipments enable transportation arrangements to be changed with an alternate route or method of transport amid delays, ensuring seamless delivery	
	Availability of backups	When the primary warehouse is down, backup facilities take over and continue the distribution process	
	Developing an emergency distribution plan	To organize an emergency delivery, the emergency supply chain (e.g., healthcare, food supply chain) can collaborate with the commercial supply chain	

(continued)

Table 1 (continued)

SC risks	Resilience strategies	Risk management application to make current SC more resilient	References
	The Omnichannel	By changing distribution pathways, the omnichannel helps to maintain material flow	
Information management risks	Blockchain technology and advanced tracking	Increases supply chain visibility, identifies problems, and aids in recovery	Durach et al. (2021), Rahman et al. (2021a, b), Ivanov (2017), Ishfaq et al. (2021)
	Implementing enterprise resource planning (ERP)	Bringing internal and external SCs together improves visibility	
	Using big data for analytics	Big data analysis in supply chains may be used for continuous monitoring, risk assessment, and opportunity mapping	
	Creating a digital twin	A cyber-physical system allows for the creation of a virtual model of a physical supply chain, forecasting, and design change	
Financial management risks	An innovative public-private partnership	During supply chain interruptions, government assistance can help alleviate financial risks	Papadopoulos et al. (2017), Ivanov and Sokolov (2019), Dong et al. (2018)
	Liquidity reserves	By maintaining a liquidity reserve, the company is able to sustain supply chain activities even in extremely disruptive situations such as the COVID-19 pandemic	
	Business insurance	If transportation is hindered or items are destroyed, insurance might act as a backup plan for the financial management of SCs	

SC methods to make SCs more robust and sustainable. Researchers have suggested that greening SCs will lead to greater sustainability, while others suggest improving the economic, social, and environmental performance of SCs (Paul et al. 2021a, b).

In times of extreme disruption, meeting customers' increased demand (due to panic-buying) helps improve the SC's social performance (Ghosh and Shah 2015). Increased manufacturing capacity ensures that customers' needs are met (Rahman et al. 2021a, b). Addressing health and safety concerns throughout SCs during massive disruptions such as the COVID-19 pandemic assists in boosting the SC's social performance (Ivanov 2021c). Increasing the capacity to manufacture biodegradable or organic products helps to improve the SC's environmental performance (Vilarinho et al. 2018). The circular economy, effective logistics, and the development of waste management capability may all help to improve SC's environmental performance (Pivnenko et al. 2016). SC's environmental challenges are sustained by green production capabilities (Hsu et al. 2013). Controlling carbon emissions improves environmental sustainability across the SC, particularly in the transportation sector (Aldrighetti et al. 2021). Checking shortage costs by swiftly satisfying consumers' demand and orders improves the SC's economic performance. Profit maximization is achieved by lowering overall SC expenses, and business diversity aids in improving the SC's economic performance (Shahed et al. 2021). Increased sharing of resources through vertical and horizontal collaboration aids in the preservation of economic performance during major interruptions, such as the COVID-19 pandemic (Mehrotra et al. 2020). Even in the face of large disruptions such as the COVID-19 pandemic, pooling financial resources among SCs and other horizontal organizations aids in the preservation of economic performance (Pettit et al. 2019). Super disruptions, such as the COVID-19 pandemic, cause SCs to be disrupted, resulting in the partial or complete closure of manufacturing facilities for a period of time. Taking urgent steps to restore SC activity by implementing recovery strategies can assist SC networks and businesses to survive economic interruptions.

Adaptation strategies, such as intertwining, substitution, scalability, and repurposing segmented by Ivanov (2020), can be used to reshape SCs when there has been disruption on a large scale in order to restore SCs to a new normal state. The following presents a brief description of the adaptation strategies:

Intertwining adaptation strategy: The COVID-19 pandemic caused severe disruptions in demand for many essential and luxury products. For example, low demand from consumers initially led to a severe shortage of semiconductors in the automotive and electronic industries. In order to accelerate their production, semiconductor companies needed to collaborate with each other; this is known as intertwining SCs (Ivanov 2021c). Manufacturers need to adapt dynamic intertwining strategies to survive disruption and ensure SCs remain viable (Salama and McGarvey 2021).

Substitution adaptation strategy: Strategies for substitution adaptation are implemented at the level of a viable SC's network and resource capabilities. Reconfiguration of network structures and product substitution are among the main adaptation strategies in this category (Ivanov 2021c). The emerging global pandemic caused by COVID-19 forced most countries of the world to close their borders to countries more susceptible to the disease (Michel-Villarreal et al. 2021). In one country, manufacturers who depended on both local and overseas suppliers were faced with severe supply shortages. For the smooth delivery of raw materials, most researchers suggested looking for alternative/backup suppliers (Paul et al. 2021a, b).

Scalability adaptation strategy: Similar to the substitution adaptation strategy, the scalability adaptation strategy works at the network and resource capabilities level of a viable SC (Ivanov 2021c). The demand for essential items, such as food, facemasks, ventilators, etc., soared during the pandemic. For essential manufacturers to fulfill the excess demand of the customers, they need to expand their production capacity and SC networks (Mohammed et al. 2021).

Repurposing adaptation strategy: The adaptive repurposing strategy also contributes to the network and resource capacity of a viable SC as well as substitution and scalability (Ivanov 2021a). In response to the pandemic, Ford Motor Company strategically used its production line to produce personal protective equipment, such as face shields (Belhadi et al. 2021). Many garment factories were unable to sell enough apparel items during the pandemic so they turned their production lines to facemask production instead (Paul et al. 2021a, b).

7 Tools, Techniques and Approaches Applied in Supply Chain Risk Management

There are three types of methodology for SC risk management approaches found in the literature: (*i*) quantitative, (*ii*) qualitative, and (*iii*) empirical methods (Chowdhury et al. 2021). Some are individual methods, while others are integrated methods. Figure 2 shows the types of SC risk management methods.

7.1 Quantitative Methods

Researchers used a range of quantitative modeling approaches to justify strategies for making SC networks more stable, viable, and sustainable. From the literature,

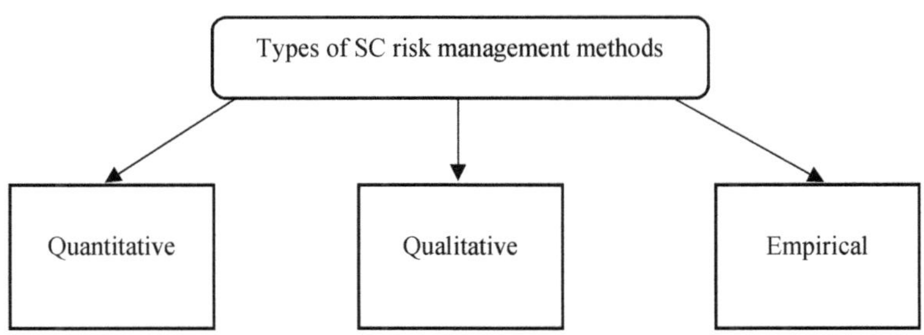

Fig. 2 Types of SC risk management methods (Rahman et al. 2022)

Ivanov and Dolgui (2021a, b) categorized modeling approaches to aid in network-wide assessing, decision-making, and process management, thus justifying steps to make SCs more resilient and sustainable.

Bayesian networks, reliability theory, complexity theory, petri nets, Markov chains, and other tools may be used to perform network-wide analyzes to identify bottlenecks in SC networks. Hosseini et al. (2019) used Bayesian networks and proposed resilience strategies for supplier selection during times of disruption. Siva Kumar and Anbanandam (2020) proposed a method based on complexity theory in order to see how the SC resilience can improve building capabilities and resilience. Gao (2015) investigated versatile risk mitigation in SCs by presenting a Markov model of demand volatility and incorporating it into dynamic mathematical programming for hedging of inventory. It is highly crucial to detect the bottlenecks in SCs that might enhance SC hazards through a network-wide investigation.

Mathematical optimization is an excellent modeling technique for making decisions. Soren and Shastri (2019) employed a multi-objective optimization approach to lower overall SC costs by managing production gaps. The model required considering disruption during procurement and production. A mathematical model developed by Lücker et al. (2019) was used to assist in analyzing decisions for appropriate inventory and redundancies under stochastic demand. Hosseini et al. (2019) evaluated resilient supplier selection approaches for recovering from supply-side disruptions using a mixed stochastic bi-objective mixed-integer programming model and a probabilistic graphical model. Tucker et al. (2020) investigated ways to reduce product (drug) shortages, and Fattahi et al. (2017) evaluated demand volatility and long lead time. Both authors examined the defined problems using multi-stage stochastic programming.

In contrast, several simulation methodologies are used for process control analysis (Ivanov and Dolgui 2021a). Using a system dynamics simulation, Chen et al. (2020) investigated the resilience initiatives for oil imports in the face of shock. A model based on agent-based simulation was developed by Rahman et al. (2021a, b) to anticipate and mitigate the impacts of the COVID-19 pandemic. However, Tan et al. (2020) examined SC resilience techniques in an SC network using a discrete event model and agent-based modeling (ABM). In exploring process control analysis, ABM is a superior method to understand stakeholder behavior in SCs. Mixed approaches of mathematical optimization and simulation methods may greatly assist in developing models for viable SC networks and evaluating process decision-making strategies for a better understanding of the consequences of large-scale SC disruptions.

7.2 Qualitative Methods

Several qualitative methods are described in the literature, either as individual methods or integrated with other qualitative or quantitative methods. Among the important qualitative methods are the case study, interview, and survey methods. For instance, Herold et al. (2021) conducted interviews with logistics service providers to learn how they dealt with the COVID-19 epidemic. Modgil et al. (2021) conducted expert interviews to investigate the prospects for improving SC resilience through distribution, visibility, and sourcing using AI. Also, Werner et al. (2021) investigated non-financial variables of organizational effectiveness in a case study. Papadopoulos et al. (2017) collected data and surveyed social media to evaluate resilience in SC networks for sustainability. In order to grasp knowledge regarding SC risk management, qualitative methods, such as case studies, surveys, and interviews, are extremely useful.

7.3 Empirical Methods

The empirical method refers to a procedure for conducting an investigation based primarily on experimentation and systematic observation rather than theoretical speculation. An empirical approach can provide valuable insight into the behavioral side of SC risk management. Empirical methods include some tools such as structural equation modeling (SEM), partial least squares-SEM (PLS-SEM), exploratory factor analysis, hierarchical regression analysis, and others. SEM was used, for example, by Vanpoucke and Ellis (2019) to examine supply-side risk management strategies. To analyze logistics resilience strategies, Liu and Lee (2018) used PLS-SEM. Using covariance-based SEM (CB-SEM), Singh and Singh (2019) examined how data analytics can help firms to make their SCs more resilient. Asamoah et al. (2020) utilized a method called exploratory factor analysis (CFA) to investigate the association between social network ties, SC resilience, and customer-oriented performance. The use of empirical methods in SC research requires primary evidence-based data where the output is reliable and applicable to the management of SC risks.

8 Technological Approaches Applied in Supply Chain Risk Management

The literature suggests several technologies for managing SC risks and disruptions. A few of the many technological approaches include blockchain analysis, big data analytics, digital twins, 3-D printing, industrial revolution 4.0, robotics, artificial intelligence, machine learning or reinforcement learning, and the Internet of Things (IoT). Supply chains can be greatly improved by blockchain technology which can offer faster and more cost-efficient delivery of products, improved traceability, improved coordination between partners, and improved access to financing. By tracing the origins of raw materials to the end consumers, including the production stages, blockchain technology can significantly assist in minimizing SC risks. Big data analytics provide the decision-makers for the SCs of companies with information and quantitative approaches that help them make more intelligent decisions.

The system consists of two new features in particular. To begin with, it expands the dataset for analysis beyond the conventional internal data held in enterprise resource planning (ERP) and SC management systems. In addition, it analyzes both new and old data sources using powerful statistical approaches. Using this method, SC decision-makers can make better decisions about everything from front-line operations to choosing the right SC operating model. A virtual SC replica with hundreds of assets, warehouses, logistics, and inventory placements is known as a digital twin. The digital twin simulates the SC's performance using advanced analytics and artificial intelligence, including all the complexity that leads to value loss and risk. In the SC, 3D printing allows for a great deal of versatility in terms of what may be printed. They print with a variety of materials, such as polymers, to print hard objects such as eyewear. By combining a rubber/plastic powder, users can create flexible items such as phone cases. During extreme disruptions like the COVID-19 pandemic, 3-D printing technology can help to add value to traditional SCs. During the pandemic, 3-D printing technology greatly aided in the production of healthcare products, such as personal protective equipment (PPE) and facemasks, to fulfill increased customer demand. Industry 4.0 necessitates a paradigm shift in how products and services are manufactured, distributed/supplied, sold, and consumed in the SC, resulting in major structural theoretical evolution and revolution in operations and SC management. Industry 4.0 is the future of SCs allowing them to deal with future disruption. Automated SC functions can be greatly streamlined by integrating robots with other automation systems. The use of IoT, artificial intelligence, machine learning, or reinforcement learning is an extremely powerful tool in capturing the big picture of SCs and solving their risk-related problems. Therefore, to manage SC risks, it is crucial for SC decision-makers to integrate the latest technologies into their SCs as much as possible.

9 Conclusions and Research Directions

A business decision-maker must act promptly in order to ensure SCs are viable in the long term, resilient, and sustainable. To ensure that SCs remain strong and sustainable, they should use reconfiguration methodologies to align SCs whenever disruptive events occur. It is imperative that SC managers develop methodologies that are reconfigurable, adaptive, robust, long-lasting, and dynamic in order to recover from super disruptions. In addition, it is crucial to understand the risks and vulnerabilities associated with SCs, their causes and factors, and how these risks and vulnerabilities can disrupt operations. This is an extremely important research avenue for SC risk management. When massive disruptions occur, such as the COVID-19 pandemic, different levels of SC networks are affected simultaneously. However, the long-term consequences remain unknown. Scaling, substitution, repurposing, and interweaving are just a few of the adaptation tactics that can be used to deal with the unknown unknowns of a pandemic. Detecting and quantifying SC network uncertainty is crucial. However, it is important to know what risks are associated with large-scale

SC disruption and how such disruption may impact operations in the long term. This is another extremely important research avenue for SC risk management.

The profitability of SC networks depends on a reconfigurable technique that reduces the uncertainty caused by large-scale outages. Simply put, companies (and governments) throughout the world should build resilience into every important SC on the globe; this is the only way to deal with large-scale unpredictability. The manufacturing SC has been underrepresented in recent SC advancements, but it should now be the focus, brought under more systematic and coordinated supervision. This is because supply-side interruptions are the most common. Many SCs were impacted by the COVID-19 pandemic, first on the supply side and then on the demand side. Rather than focusing just on compliance, procurement should begin segmenting their entire supplier pool based on capabilities, expectations, and other situations. As a result, a secure, immutable, and trustworthy channel for real-time information flow between SC stakeholders is required, which can be established using emerging technologies such as the Internet of Things (IoT), big data analytics, blockchain, reinforcement learning, and artificial intelligence. These technologies are capable of identifying the source of disturbances and providing remedies. Practitioners may use such technologies to provide reconfigurable solutions that, in many circumstances, improve SC network resilience and sustainability. They may put technologies to the test and make substantial adjustments to the strategy's dynamic to make SC networks more feasible and resilient. A better understanding of resilience and sustainability strategies, tools, techniques, modeling methods, technologies, and future business models is an important research direction for managing short- and long-term risks and disruptions to SCs.

References

Aldrighetti R, Battini D, Ivanov D, Zennaro I (2021) Costs of resilience and disruptions in supply chain network design models: a review and future research directions. Int J Prod Econ 235:108103. https://doi.org/10.1016/j.ijpe.2021.108103

Asamoah D, Agyei-Owusu B, Ashun E (2020) Social network relationship, supply chain resilience and customer-oriented performance of small and medium enterprises in a developing economy. Benchmarking 27(5):1793–1813. https://doi.org/10.1108/BIJ-08-2019-0374

Belhadi A, Kamble S, Jabbour CJC, Gunasekaran A, Ndubisi NO, Venkatesh M (2021) Manufacturing and service supply chain resilience to the COVID-19 outbreak: lessons learned from the automobile and airline industries. Technol Forecast Soc Chang 163:120447. https://doi.org/10.1016/j.techfore.2020.120447

Blackhurst J, Rungtusanatham MJ, Scheibe K, Ambulkar S (2018) Supply chain vulnerability assessment: a network based visualization and clustering analysis approach. J Purch Supply Manag 24(1):21–30. https://doi.org/10.1016/j.pursup.2017.10.004

Can Saglam Y, Yildiz Çankaya S, Sezen B (2020) Proactive risk mitigation strategies and supply chain risk management performance: an empirical analysis for manufacturing firms in Turkey. J Manuf Technol Manag 32(6):1224–1244. https://doi.org/10.1108/JMTM-08-2019-0299

Cavalcante IM, Frazzon EM, Forcellini FA, Ivanov D (2019) A supervised machine learning approach to data-driven simulation of resilient supplier selection in digital manufacturing. Int J Inf Manag 49:86–97. https://doi.org/10.1016/j.ijinfomgt.2019.03.004

Chen S, Zhang M, Ding Y, Nie R (2020) Resilience of China's oil import system under external shocks: a system dynamics simulation analysis. Energy Policy 146:111795. https://doi.org/10.1016/j.enpol.2020.111795

Choi TM (2020) Innovative "Bring-Service-Near-Your-Home" operations under Corona-Virus (COVID-19/SARS-CoV-2) outbreak: can logistics become the Messiah? Transp Res Part E: Logist Transp Rev 140:101961. https://doi.org/10.1016/j.tre.2020.101961

Chowdhury P, Kumar Paul S, Kaisar S, Abdul Moktadir M (2021) COVID-19 pandemic related supply chain studies: a systematic review. Transp Res Part E: Logist Transp Rev 148:102271. https://doi.org/10.1016/j.tre.2021.102271

Christopher M, Mena C, Khan O, Yurt O (2011) Approaches to managing global sourcing risk. Supply Chain Manag: Int J 16:67–81. https://doi.org/10.1108/13598541111115338

Diabat A, Khodaverdi R, Olfat L (2013) An exploration of green supply chain practices and performances in an automotive industry. Int J Adv Manuf Technol 68:949–961. https://doi.org/10.1007/s00170-013-4955-4

Dohale V, Verma P, Gunasekaran A, Ambilkar P (2021) COVID-19 and supply chain risk mitigation: a case study from India. Int J Logist Manag. https://doi.org/10.1108/IJLM-04-2021-0197

Dolgui A, Ivanov D (2020) Exploring supply chain structural dynamics: new disruptive technologies and disruption risks. Int J Prod Econ 229:107886. https://doi.org/10.1016/j.ijpe.2020.107886

Dolgui A, Ivanov D (2021) Ripple effect and supply chain disruption management: new trends and research directions. Int J Prod Res 59(1):102–109. https://doi.org/10.1080/00207543.2021.1840148

Dolgui A, Ivanov D, Sokolov B (2018) Ripple effect in the supply chain: an analysis and recent literature. Int J Prod Res 56(1–2):414–430. https://doi.org/10.1080/00207543.2017.1387680

Dong L, Tang SY, Tomlin B (2018) Production chain disruptions: inventory, preparedness, and insurance. Prod Oper Manag 27(7):1251–1270. https://doi.org/10.1111/poms.12866

Dubey R, Gunasekaran A, Childe SJ, Papadopoulos T, Blome C, Luo Z (2019) Antecedents of resilient supply chains: an empirical study. IEEE Trans Eng Manag 66(1):8–19. https://doi.org/10.1109/TEM.2017.2723042

DuHadway S, Carnovale S, Hazen B (2019) Understanding risk management for intentional supply chain disruptions: risk detection, risk mitigation, and risk recovery. Ann Oper Res 283:179–198. https://doi.org/10.1007/s10479-017-2452-0

Durach CF, Blesik T, von Düring M, Bick M (2021) Blockchain applications in supply chain transactions. J Bus Logist 42(1):7–24. https://doi.org/10.1111/jbl.12238

Fattahi M, Govindan K, Keyvanshokooh E (2017) Responsive and resilient supply chain network design under operational and disruption risks with delivery lead-time sensitive customers. Transp Res Part E: Logist Transp Rev 101:176–200. https://doi.org/10.1016/j.tre.2017.02.004

Gao L (2015) Collaborative forecasting, inventory hedging and contract coordination in dynamic supply risk management. Eur J Oper Res 245(1):133–145. https://doi.org/10.1016/j.ejor.2015.02.048

Gholami-Zanjani SM, Jabalameli MS, Klibi W, Pishvaee MS (2021) A robust location-inventory model for food supply chains operating under disruptions with ripple effects. Int J Prod Res 59(1):301–324. https://doi.org/10.1080/00207543.2020.1834159

Ghosh D, Shah J (2015) Supply chain analysis under green sensitive consumer demand and cost sharing contract. Int J Prod Econ 164:319–329. https://doi.org/10.1016/j.ijpe.2014.11.005

Gunasekaran A, Subramanian N, Rahman S (2015) Green supply chain collaboration and incentives: current trends and future directions. Transp Res Part E: Logist Transp Rev 74:1–10. https://doi.org/10.1016/j.tre.2015.01.002

Gupta V, Ivanov D (2020) Dual sourcing under supply disruption with risk-averse suppliers in the sharing economy. Int J Prod Res 58(1):291–307. https://doi.org/10.1080/00207543.2019.1686189

Gurtu A, Johny J (2021) Supply chain risk management: literature review. Risks 9(1):1–16. https://doi.org/10.3390/risks9010016

Hall MC, Prayag G, Fieger P, Dyason D (2020) Beyond panic buying: consumption displacement and COVID-19. J Serv Manag 32(1):113–128. https://doi.org/10.1108/JOSM-05-2020-0151
Herold DM, Nowicka K, Pluta-Zaremba A, Kummer S (2021) COVID-19 and the pursuit of supply chain resilience: reactions and "lessons learned" from logistics service providers (LSPs). Supply Chain Manag 26(6):702–714. https://doi.org/10.1108/SCM-09-2020-0439
Ho W, Zheng T, Yildiz H, Talluri S (2015) Supply chain risk management: a literature review. Int J Prod Res. https://doi.org/10.1080/00207543.2015.1030467
Hobbs JE (2020) Food supply chains during the COVID-19 pandemic. Can J Agric Econ 68(2):171–176. https://doi.org/10.1111/cjag.12237
Hosseini S, Morshedlou N, Ivanov D, Sarder MD, Barker K, Khaled AA (2019) Resilient supplier selection and optimal order allocation under disruption risks. Int J Prod Econ 213:124–137. https://doi.org/10.1016/j.ijpe.2019.03.018
Hsu C, Choon Tan K, Zailani HM, S., & Jayaraman, V. (2013) Supply chain drivers that foster the development of green initiatives in an emerging economy. Int J Oper Prod Manag 33(6):656–688. https://doi.org/10.1108/IJOPM-10-2011-0401
Ishfaq R, Davis-Sramek E, Gibson B (2021) Digital supply chains in omnichannel retail: a conceptual framework. J Bus Logist. https://doi.org/10.1111/jbl.12277
Ivanov D (2017) Simulation-based single vs. dual sourcing analysis in the supply chain with consideration of capacity disruptions, big data and demand patterns. Int J Integr Supply Manag 11(1):24–43. https://doi.org/10.1504/IJISM.2017.083005
Ivanov D (2019) Disruption tails and revival policies: a simulation analysis of supply chain design and production-ordering systems in the recovery and post-disruption periods. Comput Ind Eng 127:558–570. https://doi.org/10.1016/j.cie.2018.10.043
Ivanov D (2020) Predicting the impacts of epidemic outbreaks on global supply chains: a simulation-based analysis on the coronavirus outbreak (COVID-19/SARS-CoV-2) case. Transp Res Part E: Logist Transp Rev 136:101922. https://doi.org/10.1016/j.tre.2020.101922
Ivanov D (2021a) Exiting the COVID-19 pandemic: after-shock risks and avoidance of disruption tails in supply chains. Ann Oper Res. https://doi.org/10.1007/s10479-021-04047-7
Ivanov D (2021b) Lean resilience: AURA (Active usage of resilience assets) framework for post-COVID-19 supply chain management. Int J Logist Manag. https://doi.org/10.1108/IJLM-11-2020-0448
Ivanov D (2021c) Supply Chain Viability and the COVID-19 pandemic: a conceptual and formal generalisation of four major adaptation strategies. Int J Prod Res 59(12):3535–3552. https://doi.org/10.1080/00207543.2021.1890852
Ivanov D, Dolgui A (2021a) OR-methods for coping with the ripple effect in supply chains during COVID-19 pandemic: managerial insights and research implications. Int J Prod Econ 232:107921. https://doi.org/10.1016/j.ijpe.2020.107921
Ivanov D, Dolgui A (2021b) Stress testing supply chains and creating viable ecosystems. Oper Manag Res. https://doi.org/10.1007/s12063-021-00194-z
Ivanov D, Sokolov B (2010) Adaptive supply chain management. Adapt Supply Chain Manag. https://doi.org/10.1007/978-1-84882-952-7
Ivanov D, Sokolov B (2019) Simultaneous structural—Operational control of supply chain dynamics and resilience. Ann Oper Res 283(1–2):1191–1210. https://doi.org/10.1007/s10479-019-03231-0
Kamalahmadi M, Parast MM (2017) An assessment of supply chain disruption mitigation strategies. Int J Prod Econ 184:210–230. https://doi.org/10.1016/j.ijpe.2016.12.011
Karmaker CL, Ahmed T, Ahmed S, Ali SM, Moktadir MA, Kabir G (2021) Improving supply chain sustainability in the context of COVID-19 pandemic in an emerging economy: exploring drivers using an integrated model. Sustain Prod Consum 26:411–427. https://doi.org/10.1016/j.spc.2020.09.019
Kilubi I (2016) The strategies of supply chain risk management—A synthesis and classification. Int J Log Res Appl 19(6):604–629. https://doi.org/10.1080/13675567.2016.1150440

Lintukangas K, Kähkönen AK, Ritala P (2016) Supply risks as drivers of green supply management adoption. J Clean Prod 112:1901–1909. https://doi.org/10.1016/j.jclepro.2014.10.089

Liu CL, Lee MY (2018) Integration, supply chain resilience, and service performance in third-party logistics providers. Int J Logist Manag 29(1):5–21. https://doi.org/10.1108/IJLM-11-2016-0283

Lücker F, Seifert RW, Biçer I (2019) Roles of inventory and reserve capacity in mitigating supply chain disruption risk. Int J Prod Res 57(4):1238–1249. https://doi.org/10.1080/00207543.2018.1504173

Luthra S, Kumar V, Kumar S, Haleem A (2011) Barriers to implement green supply chain management in automobile industry using interpretive structural modeling technique-an Indian perspective. J Ind Eng Manag 4(2):231–257. https://doi.org/10.3926/jiem.2011.v4n2.p231-257

Macdonald JR, Zobel CW, Melnyk SA, Griffis SE (2018) Supply chain risk and resilience: theory building through structured experiments and simulation. Int J Prod Res 56(12):4337–4355. https://doi.org/10.1080/00207543.2017.1421787

Majumdar A, Sinha SK, Govindan K (2021) Prioritising risk mitigation strategies for environmentally sustainable clothing supply chains: insights from selected organisational theories. Sustain Prod Consum 28:543–555. https://doi.org/10.1016/j.spc.2021.06.021

Manuj I, Esper TL, Stank TP (2014) Supply chain risk management approaches under different conditions of risk. J Bus Logist 35(3):241–258. https://doi.org/10.1111/jbl.12051

Mehrotra S, Rahimian H, Barah M, Luo F, Schantz K (2020) A model of supply-chain decisions for resource sharing with an application to ventilator allocation to combat COVID-19. Nav Res Logist 67(5):303–320. https://doi.org/10.1002/nav.21905

Michel-Villarreal R, Vilalta-Perdomo EL, Canavari M, Hingley M (2021) Resilience and digitalization in short food supply chains: a case study approach. Sustainability (switzerland) 13(11):1–23. https://doi.org/10.3390/su13115913

Modgil S, Singh RK, Hannibal C (2021) Artificial intelligence for supply chain resilience: learning from Covid-19. Int J Logist Manag. https://doi.org/10.1108/IJLM-02-2021-0094

Mohammed A, Naghshineh B, Spiegler V, Carvalho H (2021) Conceptualising a supply and demand resilience methodology: a hybrid DEMATEL-TOPSIS-possibilistic multi-objective optimization approach. Comput Ind Eng 160:107589. https://doi.org/10.1016/j.cie.2021.107589

Moosavi J, Hosseini S (2021) Simulation-based assessment of supply chain resilience with consideration of recovery strategies in the COVID-19 pandemic context. Comput Ind Eng 160:107593. https://doi.org/10.1016/j.cie.2021.107593

Mu W, van Asselt ED, van der Fels-Klerx HJ (2021) Towards a resilient food supply chain in the context of food safety. Food Control 125:107953. https://doi.org/10.1016/j.foodcont.2021.107953

Nicola M, Alsafi Z, Sohrabi C, Kerwan A, Al-Jabir A, Iosifidis C, Agha M, Agha R (2020) The socio-economic implications of the coronavirus pandemic (COVID-19): a review. Int J Surg 78:185–193. https://doi.org/10.1016/j.ijsu.2020.04.018

Papadopoulos T, Gunasekaran A, Dubey R, Altay N, Childe SJ, Fosso-Wamba S (2017) The role of big data in explaining disaster resilience in supply chains for sustainability. J Clean Prod 142(2):1108–1118. https://doi.org/10.1016/j.jclepro.2016.03.059

Paul SK, Chowdhury P (2020a) A production recovery plan in manufacturing supply chains for a high-demand item during COVID-19. Int J Phys Distrib Logist Manag 51:101–125. https://doi.org/10.1108/IJPDLM-04-2020-0127

Paul SK, Chowdhury P (2020b) Strategies for managing the impacts of disruptions during COVID-19: an example of toilet paper. Glob J Flex Syst Manag. https://doi.org/10.1007/s40171-020-00248-4

Paul SK, Chowdhury P, Moktadir A, Lau KH (2021a) Supply chain recovery challenges in the wake of COVID-19 pandemic. J Bus Res 136:316–329. https://doi.org/10.1016/j.jbusres.2021.07.056

Paul SK, Moktadir MA, Sallam K, Choi TM, Chakrabortty RK (2021b) A recovery planning model for online business operations under the COVID-19 outbreak. Int J Prod Res. https://doi.org/10.1080/00207543.2021.1976431

Paul SK, Sarker R, Essam D (2017) A quantitative model for disruption mitigation in a supply chain. Eur J Oper Res 257(3):881–895. https://doi.org/10.1016/j.ejor.2016.08.035

Pavlov A, Ivanov D, Werner F, Dolgui A, Sokolov B (2019) Integrated detection of disruption scenarios, the ripple effect dispersal and recovery paths in supply chains. Ann Oper Res. https://doi.org/10.1007/s10479-019-03454-1

Pettit TJ, Croxton KL, Fiksel J (2019) The evolution of resilience in supply chain management: a retrospective on ensuring supply chain resilience. J Bus Logist 40(1):56–65. https://doi.org/10.1111/jbl.12202

Pivnenko K, Eriksen MK, Martín-Fernández JA, Eriksson E, Astrup TF (2016) Recycling of plastic waste: presence of phthalates in plastics from households and industry. Waste Manag 54:44–52. https://doi.org/10.1016/j.wasman.2016.05.014

Pournader M, Kach A, Talluri S (2020) A review of the existing and emerging topics in the supply chain risk management literature. Decis Sci 51(4):867–919. https://doi.org/10.1111/deci.12470

Prakash S, Soni G, Rathore APS (2017) A critical analysis of supply chain risk management content: a structured literature review. J Adv Manag Res 14(1):69–90. https://doi.org/10.1108/JAMR-10-2015-0073

Rahman T, Moktadir MA, Paul SK (2021a) Key performance indicators for a sustainable recovery strategy in health-care supply chains: COVID-19 pandemic perspective. J Asia Bus Stud. https://doi.org/10.1108/JABS-05-2021-0200

Rahman T, Taghikhah F, Paul SK, Shukla N, Agarwal R (2021b) An agent-based model for supply chain recovery in the wake of the COVID-19 pandemic. Comput Ind Eng 158:107401. https://doi.org/10.1016/j.cie.2021.107401

Rahman T, Paul SK, Shukla N, Agarwal R, Taghikhah F (2022) Supply chain resilience initiatives and strategies: a systematic review. Comput Ind Eng 170:108317. https://doi.org/10.1016/j.cie.2022.108317

Razavian E, Alem Tabriz A, Zandieh M, Hamidizadeh MR (2021) An integrated material-financial risk-averse resilient supply chain model with a real-world application. Comput Ind Eng 161:107629. https://doi.org/10.1016/j.cie.2021.107629

Rehman O, Ali Y (2021) Enhancing healthcare supply chain resilience: decision-making in a fuzzy environment. Int J Logist Manag. https://doi.org/10.1108/IJLM-01-2021-0004

Salama MR, McGarvey RG (2021) Resilient supply chain to a global pandemic. Int J Prod Res. https://doi.org/10.1080/00207543.2021.1946726

Scala B, Lindsay CF (2021) Supply chain resilience during pandemic disruption: evidence from healthcare. Supply Chain Manag 26(6):672–688. https://doi.org/10.1108/SCM-09-2020-0434

Shahed KS, Azeem A, Ali SM, Moktadir MA (2021) A supply chain disruption risk mitigation model to manage COVID-19 pandemic risk. Environ Sci Pollut Res. https://doi.org/10.1007/s11356-020-12289-4

Shekarian M, Mellat Parast M (2021) An Integrative approach to supply chain disruption risk and resilience management: a literature review. Int J Log Res Appl 24(5):427–455. https://doi.org/10.1080/13675567.2020.1763935

Shishodia A, Sharma R, Rajesh R, Munim ZH (2021) Supply chain resilience: a review, conceptual framework and future research. Int J Logist Manag. https://doi.org/10.1108/IJLM-03-2021-0169

Singh NP, Singh S (2019) Building supply chain risk resilience: role of big data analytics in supply chain disruption mitigation. Benchmarking 26(7):2318–2342. https://doi.org/10.1108/BIJ-10-2018-0346

Siva Kumar P, Anbanandam R (2020) Theory building on supply chain resilience: A SAP-LAP analysis. Glob J Flex Syst Manag 21:113–133. https://doi.org/10.1007/s40171-020-00233-x

Soren A, Shastri Y (2019) Resilient design of biomass to energy system considering uncertainty in biomass supply. Comput Chem Eng 131:106593. https://doi.org/10.1016/j.compchemeng.2019.106593

Tan WJ, Cai W, Zhang AN (2020) Structural-aware simulation analysis of supply chain resilience. Int J Prod Res 58(17):5175–5195. https://doi.org/10.1080/00207543.2019.1705421

Tarafdar M, Qrunfleh S (2017) Agile supply chain strategy and supply chain performance: complementary roles of supply chain practices and information systems capability for agility. Int J Prod Res 55(4):925–938. https://doi.org/10.1080/00207543.2016.1203079

Tucker EL, Daskin MS, Sweet BV, Hopp WJ (2020) Incentivizing resilient supply chain design to prevent drug shortages: policy analysis using two- and multi-stage stochastic programs. IISE Trans 52(4):394–412. https://doi.org/10.1080/24725854.2019.1646441

Valipour Parkouhi S, Safaei Ghadikolaei A (2017) A resilience approach for supplier selection: using fuzzy analytic network process and grey VIKOR techniques. J Clean Prod 161:431–451. https://doi.org/10.1016/j.jclepro.2017.04.175

Vanpoucke E, Ellis SC (2019) Building supply-side resilience—A behavioural view. Int J Oper Prod Manag 40(1):11–33. https://doi.org/10.1108/IJOPM-09-2017-0562

Vilarinho F, Sanches Silva A, Vaz MF, Farinha JP (2018) Nanocellulose in green food packaging. Crit Rev Food Sci Nutr 58(9):1526–1537. https://doi.org/10.1080/10408398.2016.1270254

Wang M, Yao J (2021) Intertwined supply network design under facility and transportation disruption from the viability perspective. Int J Prod Res. https://doi.org/10.1080/00207543.2021.1930237

Werner MJE, Yamada APL, Domingos EGN, Leite LR, Pereira CR (2021) Exploring organizational resilience through key performance indicators. J Ind Prod Eng 38(1):51–65. https://doi.org/10.1080/21681015.2020.1839582

Wildgoose N (2016) Supply chain risk management. In: Enterprise risk management: a common framework for the entire organization. https://doi.org/10.1016/B978-0-12-800633-7.00006-7

A Review on Uncertainty Modeling for Decentralized Supply Chain Systems

Marjia Haque, Sanjoy Kumar Paul, Ruhul Sarker, and Daryl Essam

1 Introduction

Over the past few decades, supply chain (SC) planning has received substantial attention from both academics and practitioners to achieve successful business operations. A SC can be defined as an integrated network consisting of a set of entities (organizations or individuals) involved in the upstream and downstream flows of products, services, finances, and/or information from a source to a customer. Subsequently, supply chain management (SCM) is the systemic, strategic coordination of various business functions and strategies within a particular company and/or across other business entities within a network, to improve the performance of the individual companies and the chain as a whole (Mentzer et al. 2001). It is a challenging and complex decision-making process to ensure coordination across a whole chain consisting of multiple entities with several activities. However, depending on how decisions are made, SCs can be classified into two major types (Duan and Warren Liao 2013): (a) Centralized (decisions are made centrally by considering all members together) (b) Decentralized (each member makes decisions without considering others).

M. Haque (✉) · R. Sarker · D. Essam
School of Engineering and Information Technology, University of New South Wales, Canberra, Australia
e-mail: marjia.haque@adfa.edu.au

R. Sarker
e-mail: r.sarker@adfa.edu.au

D. Essam
e-mail: d.essam@adfa.edu.au

S. K. Paul
UTS Business School, University of Technology Sydney, Sydney, Australia
e-mail: Sanjoy.Paul@uts.edu.au

S. K. Paul et al. (eds.), *Supply Chain Risk and Disruption Management*, Flexible Systems Management, https://doi.org/10.1007/978-981-99-2629-9_2

In the literature, most of the SC studies considered centralized structures assuming a 'global organizer' for managing 'all' the activities of SC networks Gao and You (2018), Roghanian et al. (2007). However, optimizing a total SC using some combined or central objectives and constraints under a centralized scenario is not common in most realistic scenarios. This type of centralized structure usually exists in vertically integrated organizations that require a high degree of collaboration among their members (Haque et al. 2020). It is possible to share all the information among them to make centralized decisions. Although an integrated centralized solution may result in optimal system performance for a multi-stage SC, the solution is not always useful for every member of the chain. In practice, most organizations are separated and operate as independent economic entities (Li and Wang 2007). Hence, this type of decision-making leads to a distributed, decentralized decision-making structure, which is the reality for many SCs. Decentralized decision-making structures are widely seen in different types of business organizations; for example, supermarket and grocery SCs (Swaminathan et al. 1998), pharmaceutical chains (Nematollahi et al. 2017), retail markets, electricity markets (Zaman et al. 2017), decentralized customized manufacturing industries (Mourtzis and Doukas 2012) and many other multi-agent SC networks consisting of individual business entities and others, where suppliers, manufacturers and/or distributors of a particular product come from various organizations. Again, most organizations in the 21st-century operate under DSCs as they focus on market specialization. As such, in the healthcare industry, Philips Healthcare Netherlands, a world-known medical equipment manufacturer, has several individual manufacturing plants in Europe and North America and a major dealer in Hong Kong for fulfilling demand in Asia (Zhu 2015). Moreover, modern globalization and market specialization have forced many businesses toward decentralized production and distribution networks which can reduce their costs as well as increase their efficiency to satisfy customer demand (Sun et al. 2018). Also, many organizations have introduced outsourcing in their operational planning, which creates decentralized structures (Wu et al. 2018). In addition, assembly firms in which product parts come from other business entities operate mainly under DSCs (Gerchak and Wang 2004). Information technology has enabled many organizations to move from traditional 'vertical' integration toward today's 'virtual' integration of SC members, whereby one decision-maker (DM) coordinates with the others while sharing only necessary information without being performed as a whole. For example, the giant computer manufacturing company 'Dell' has increased its profits through its successful and innovative SC network which is distributed around the world virtually (Lawton and Michaels 2001). In recent years, third party logistics SC systems are very common, where a logistics provider may contact various manufacturers or sellers simultaneously to deliver a variety of items toward them, who are mostly belonging to different organizations and hence create a distributed network. Thus, decentralized distributed SC structures are more frequently observed these days. These individual SC members are generally more interested in optimizing their individual objectives than those of the entire system, which creates challenges in overall chain coordination (Haque et al. 2021).

In addition, uncertainty or risk is inevitable in any SC system in reality. It becomes severe in the case of decentralized structures as individual members are interested in their own decision-making process with less consideration of the uncertain parameters of others (Hjaila et al. 2016). This uncertainty can occur in any part or stage of a chain and can affect a whole system. Uncertainty over customer behaviors creates significant risk for all DSC partners in which members' decisions vary based on their attitude toward risk management (Hafezalkotob et al. 2011). In recent years, the COVID-19 outbreak has further proved SC risk and vulnerabilities across various organizations due to travel bans, factory shutdowns, lockdowns in the supply market, labor shortages, and all other uncertain and unexpected incidents (Chowdhury et al. 2021; Omar et al. 2022). Hence, firms and industries have incredibly realized the need for adaptive and resilient SC systems to sustain all future risks and uncertainties (Queiroz et al. 2020). Consequently, an increasing number of studies have recently been observed, focusing on DSC systems under uncertainties or risks using various quantitative modeling approaches, as depicted in Fig. 1.

With a significant number of research focusing on DSC modeling considering uncertainties or risks, this chapter aims to outline the current state of literature in this domain and assist the researchers in further study. The specific objectives of this chapter are as follows.

- To review different quantitative models of DSC planning under uncertainties.
- To identify existing studies on different strategies of quantitative mathematical models or approaches to coordinate independent members of a DSC under limited information-sharing.
- To identify research gaps and suggest a few research directions.

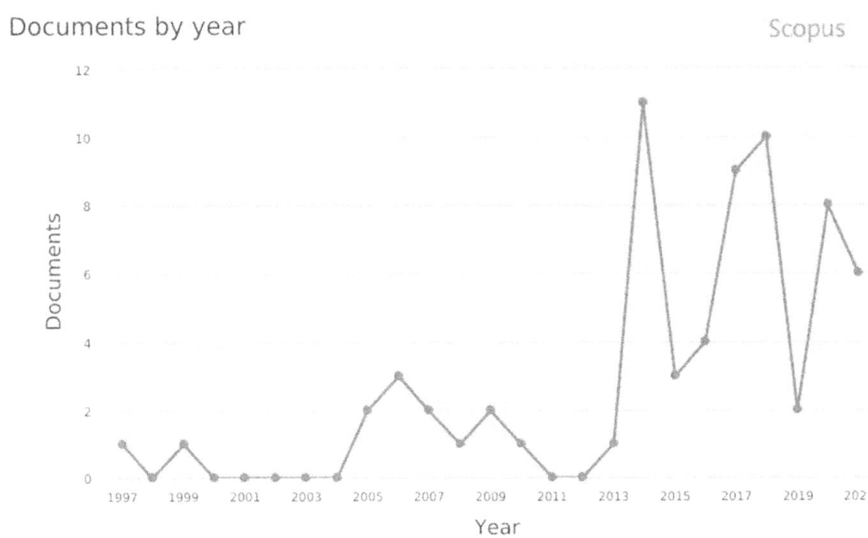

Fig. 1 Number of articles published on DSC modeling under uncertainties or risk. *Source* Scopus

The remainder of the chapter is organized as follows. After the introduction, the review methodology of this study is described in Sect. 2. Section 3 presents an analysis of the reviewed articles with DSC structures under uncertainties and/or risks using different quantitative models and various solution approaches. A summary of research gaps in this domain is discussed in Sect. 4. Finally, a few future research directions are provided, and conclusions are drawn in Sects. 5 and 6, respectively.

2 Review Methodology

A few steps have been followed in this chapter to conduct the literature review. At first, the main research topic was 'DSC modeling under uncertainties and/or risks'. Second, multiple research databases, such as Google Scholar and Scopus databases, searched for relevant articles. We considered various types of research articles, including journal papers, conference proceedings, and review papers. Next, a cross-reference check was conducted to retrieve more relevant articles for review. The inclusion criteria for our search process include articles or papers written in English and focusing on DSC quantitative modeling or optimization approaches under uncertainties and/or risks. The papers that studied centralized SC structures, non-related with modeling approaches and/or deterministic scenarios, were excluded. Figure 2 illustrates the steps followed in the review process in this chapter.

3 Analysis of Reviewed Articles

Several researchers have worked on ensuring coordination or alignment under uncertainties among different members in a chain who are not fully controlled by a single authority. They considered various problem scenarios with specific assumptions, developed models for representing those problems mathematically, and proposed various solution techniques for solving the models. This chapter will focus on a literature review of SC with decentralized decision-making strategies under uncertainties from a variety of aspects and subsequently will highlight research gaps in this field. In this chapter, the review of DSC is conducted based on three major categories: (i) different quantitative models used in DSC planning, (ii) different solution approaches used for solving those models, and (iii) applications in real-life problems. The first two categories are divided into various sub-categories. For better understanding, a detailed road map of the literature on the DSC problem reviewed in this chapter is presented in Fig. 3.

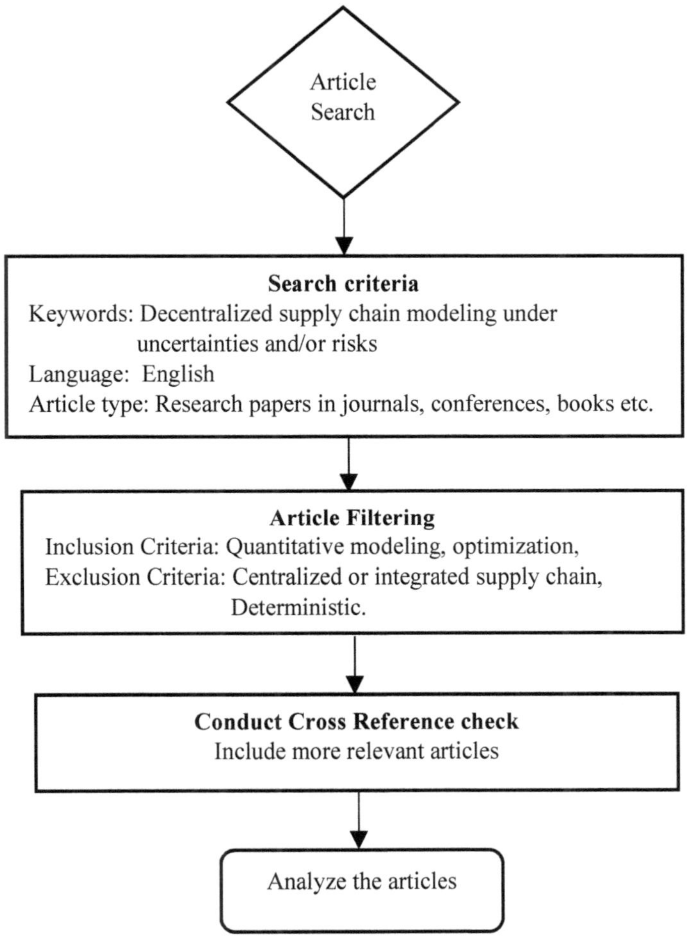

Fig. 2 Steps of review methodology

3.1 Quantitative Models

Researchers have attempted to formulate multi-stage DSCs while assuming different structures, scenarios, and information-sharing mechanisms to ensure coordination among its members under uncertain conditions. Compared with that on centralized systems, the literature on decentralized ones is less extensive. In the following subsections, quantitative models for DSC planning under various stochastic (uncertain parameters) scenarios are discussed.

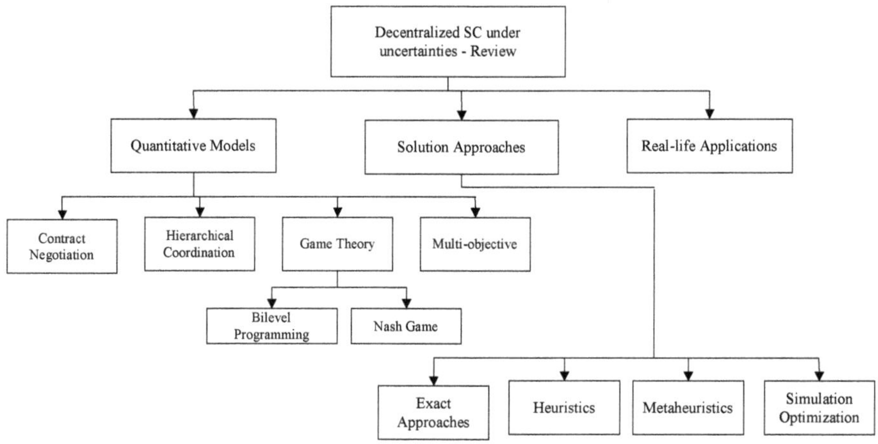

Fig. 3 Road map of reviewed research on DSC planning under uncertainties

3.1.1 Modeling with Contract Negotiation

A significant number of studies are found on DSCs that deal with contracting mechanisms among the individual members in a chain to ensure coordination under various uncertain scenarios. A contract can be defined as a set of policies applied to coordinate various SC activities, where the set of optimal actions is a Nash Equilibrium (NE), i.e., no firm has a profitable unilateral deviation from these actions (Cachon 2003). In particular, it is an agreement among SC members regarding business parameters (i.e., pricing, order quantity commitment, delivery commitment, etc.) to perform business processes (Hu et al. 2013). Among various types of contracts, buyback, revenue-sharing, and quantity discount contracts are the most studied.

Studies have considered uncertainties in SC modeling using various coordinating contracts. As such, Bernstein and Federgruen (2005) investigated the equilibrium behavior of DSCs with competing retailers and a single supplier under demand uncertainty. They designed some contractual arrangements between parties for coordination. Gerchak and Wang (2004) studied both revenue-sharing and wholesale price contracts between an assembler/retailer and its suppliers under a demand uncertainty scenario. They studied different parameters of the contracts and showed how the contracts could improve channel coordination and profit-sharing among partners. However, they pointed out the importance of considering asymmetric information about the demand forecast and/or each member's costs in the model to better represent the DSC scenarios.

He and Zhao (2012) developed a model using a returns policy combined with a wholesale price contract for a three-stage DSC consisting of supplier-manufacturer-retailer under stochastic demand and an uncertain raw material yield. Their model used a sequential flow of some SC activities among the members and was solved using a backward induction procedure. They also investigated the contract terms required to obtain optimal decisions from the perspective of the whole SC. Later, He and Zhao

(2012) and Hu et al. (2013) identified that traditional price-only contracts could not generally coordinate a DSC. Subsequently, many researchers developed models combining two or more different types of contracts to ensure coordination among independent members. For example, Hu et al. (2013) proposed a flexible ordering policy with revenue-sharing contracts to coordinate manufacturer-supplier systems while considering production yield and market demand uncertainties. Taghipour and Frayret (2013) developed a model with the operational plans of two independent SC partners using an incentive system to encourage the partners to participate in the coordination process. A contract-based sourcing strategy was also proposed by Asian and Nie (2014), considering both demand and supply uncertainties for a DSC setting to mitigate the associated risks.

Heydari and Norouzinasab (2015) studied a two-level quantity discount policy and incentive schemes for a two-echelon DSC to coordinate ordering and pricing strategies throughout a chain. They considered demand as a stochastic and price-sensitive parameter. In their model, firstly, the downstream stage decided on the selling price and order size, after which the upstream stage determined the production quantity. Fu et al. (2015) established a model for addressing the management problems of a decentralized hybrid push–pull assembly system. They used a three-stage game to analyze each player's decision problem. To ensure coordination, they proposed a buyback policy between powerful push suppliers and the assembler and a subsidy policy between the powerful assembler and pulled suppliers while considering unreliable supply and uncertain demand. A production-commitment contract was proposed by Muzaffar et al. (2017) for a manufacturer-retailer system under information asymmetry with uncertain demand. To ensure a win–win situation, they proposed a new return contract for a make-to-stock scenario in a DSC. They assumed that demand information was kept as the retailer's private information, whereas the manufacturer had to make an educated guess about the average demand and its variation (mean and standard deviation). Basu et al. (2018) proposed a buyback contract for a supplier–retailer DSC under demand uncertainty. A buyback contract was also used by Liu et al. (2018) to coordinate a manufacturer–retailer system to determine the retailer's ordering decisions under demand uncertainties. Arikan and Silbermayr (2017) proposed coordinating contracts considering risk pooling for a two-stage DSC structure. In a recent study, Giri and Sarker (2019) analyzed a DSC under uncertain price-sensitive market demand and production disruptions for both supplier and manufacturer end. They studied revenue-sharing contracts using pairwise and spanning mechanisms in the case of a multi-stage network. However, most of these studies with contract mechanisms dealt with a two-stage SC, in which contracts could be established between interacting members to ensure some level of coordination. Therefore, it is difficult to establish these types of contracting mechanisms for a multi-stage complex SC structure with multiple entities.

3.1.2 Modeling with Hierarchical Coordination

A significant number of studies of DSC modeling have been conducted using hierarchical top-down approaches for coordinating its members. In this approach, the optimized decisions of each decentralized stage are considered separately and then coordinated according to the SC's hierarchical configuration. Decision-making according to a hierarchical structure is a common way of planning for a multi-stage SC system, even under stochastic scenarios. For instance, Jalil et al. (2018) developed a multi-level model for a DSC network consisting of multiple DMs that made various decisions according to different levels, based on the hierarchical structure of the chain. They considered a fuzzy-based solution approach to handle uncertainties in the decision parameters while assuming inadequate historical data or information. Fu and Ma (2019) proposed some schemes for coordinating a two-stage DSC, connected by cross-shareholding (equity shares between vertically related firms), for both push and pull systems using the hierarchical structure of a chain. They developed models for investigating decisions regarding wholesale price and production quantities of the stages, while considering random demand and cost-demand information-sharing among the stages. According to their model, for push strategy, firstly, the upstream firm made its decisions and then a downstream one did, whereas, for the pull strategy, the decisions of the two firms were reversed. They concluded that to achieve a win–win coordination, one member should operate at a loss first and then obtain revenue by sharing its partner's profit.

3.1.3 Modeling with Game Theory

Apart from hierarchical approaches with mechanisms for changing solutions among stages, numerous researchers used game theory strategies to model DSC structures considering uncertainties. Leng and Parlar (2016) presented a detailed review of game-theoretical applications in SCM problems. They discussed both non-cooperative and cooperative games in different SC areas. The two common game theory strategies used in the literature are: Bi-level programming or the Stackelberg game approach and the Nash game approach. Both are discussed in the following sub-sections under stochastic conditions.

(a) **Bi-level Programming**

Over the last few years, a wide range of research has been conducted using different types of game theories for DSC modeling, mainly bi-level programming or Stackelberg game strategies, in which a specific stage or member is assumed as a leader (dominant) over other(s) stages acting as a follower(s). The bi-level programming problem (BLPP), or the Stackelberg game, is widely used in DSC modeling when the SC players are individual and non-cooperative. It is a special case of multi-level programming for hierarchical decision-making processes, whereby the DM at the upper level (leader) first specifies a strategy, and then the DM at the lower-level (follower) specifies a strategy to optimize their objectives, with full knowledge about the action of the DM at the upper level. Although a single-level centralized approach is more profitable as it has complete information about each member in the chain, a bi-level model is more practical for a decentralized problem, as Yeh et al. (2014) suggested.

BLPP has been widely used in SC modeling under various uncertain scenarios. Roghanian et al. (2007) considered a bi-level linear multi-objective programming task for an enterprise-wide SC planning problem, in which market demand and warehouse capacity were assumed as random variables. Hsieh and Lu (2010) proposed a pricing and ordering game model in a two-stage SC consisting of one manufacturer (Stackelberg leader) and two competing retailers under price-sensitive random demand. They used the manufacturer's return policy to coordinate the decentralized members and also investigated the effect of the asymmetry of the model's demand distribution. Xiao et al. (2010) developed a dynamic game-theoretic model for a three-stage DSC consisting of one retailer, one manufacturer, and one subcontractor under demand uncertainty. They assumed the system as a 'make-to-order' production system, where the manufacturer received the order from the retailer and then produced and delivered the ordered products to the retailer. Qiang et al. (2013) analyzed a multi-stage decentralized closed-loop SC network consisting of suppliers, manufacturers, and retail outlets under uncertainties in demand. They derived the optimal conditions for the various independent DMs and formulated the equilibrium conditions as a finite-dimensional variational inequality problem. They used a Stackelberg game, with the manufacturer as the leader and the supplier as a follower and both having symmetric information. Tang et al. (2014) used a Stackelberg game to highlight buyers' (leader) and suppliers' optimal parameter choices under stochastic demand. They presented some series of models to study the interactions between the two firms and proposed incentive mechanisms as an equilibrium solution of the game.

Nourifar et al. (2018) used a BLPP model to solve a decentralized production-distribution planning problem under uncertainties. At the upper level, they considered a distribution company to control the opening of existing distribution centers. They considered the manufacturing company to handle plants' operational costs under uncertain product demand and the customer-end price at the lower level. Hjaila et al. (2016) proposed a scenario-based dynamic negotiation approach for the coordination of DSCs under uncertainty among the client (leader), provider (follower), and third parties. According to their model, a set of coordination contracts would be prepared by the leader considering the negotiation items based on non-zero-sum non-symmetric roles, where in turn, the follower would assess the risks associated with accepting or rejecting this cooperation agreement. The evaluation would be based on the probability distribution and cumulative curve of its own expected profits, by randomly generating and simulating scenarios using a Monte Carlo approach.

A two-stage stochastic programming approach, which is popularly used to handle uncertainties in CSC structures, was also studied for decentralized cases. Gao and You (2018) developed a modeling framework that integrated the leader–follower Stackelberg game with a two-stage stochastic programming approach considering uncertainties. Also, Gao and You (2019) studied a game theory-based stochastic model to analyze the influences of uncertainties in a multi-stakeholder non-cooperative structure, following a single-leader-multiple-follower Stackelberg game scenario.

In recent studies, researchers combined the Stackelberg game theory approach with some contracting mechanisms to obtain coordination in multi-stage decentralized structures. Alamdar et al. (2019) developed stochastic Stackelberg game theory models that allowed different SC partners to be channel leaders for a closed-loop network under uncertain price and sales effort-dependent demand. They implemented a coordination contract to increase the performance of a decentralized network when compared with that of a centralized one. Later, Li et al. (2020) studied a SC with two independent agents, a manufacturer and a retailer (Stackelberg leader), while considering the product's demand and price as uncertain parameters. They proposed two types of contracts, i.e., capacity reservation and quantity flexibility, with some penalty parameters as risk-sharing mechanisms among the members to tackle this uncertainty. Glock et al. (2020) developed a model for determining the optimal order quantity and production capacity for products with a short life cycle in a manufacturer (Stackelberg leader)—retailer system while considering both demand and lead time as uncertain parameters. They proposed a buyback contract that resulted in a win–win situation for both players, in which the retailer's order quantity and the manufacturer's investments were jointly coordinated. In another study, BLPP was extended as a hierarchical Tri-Level Programming (TLP) problem by Nourifar et al. (2020) for a multi-period, multiproduct, multi-echelon SC while considering multiple uncertainties with fuzzy and stochastic parameters. They proposed an iterative solution approach assuming the distributor as the top-level leader in the hierarchical structure and the manufacturer and supplier as the middle and bottom level followers, respectively.

Although the demand is mostly considered an uncertain parameter by researchers, few studies have considered supply, price, or other parameters stochastic. For example, Schmitt et al. (2015) investigated an optimal system design in a multi-location SC with supply and demand uncertainties. Their numerical studies showed that a decentralized system was better than a centralized one when supply disruptions and demand uncertainty were present. Later, Shao et al. (2015) introduced the concept of 'SC cell' for a multi-level DSC logistics planning problem under an uncertain environment. They proposed a chance-constrained programming model that considered stochastic supply and demand between the SC nodes. Schildbach and Morari (2016) also considered various sources of uncertainties (demand, lead time, prices, etc.) and developed an approach, called scenario-based model predictive control (SCMPC), to model a multi-echelon SC.

Zhu (2015) developed a decision model for a supplier-retailer DSC system facing price and lead time dependent stochastic demand while assuming that all information was common knowledge to both players. They considered a Steckelberg game where, as a leader, the supplier made the capacity and wholesale price decisions, whereas the retailer, as a follower, determined the sales price and delivery time based on the supplier's decisions. They proposed a franchise contract to achieve channel coordination and a win–win situation as they found revenue-sharing and two-part tariff contracts were not suitable for coordinating such a chain. Yin et al. (2014) developed a Stackelberg game-theoretic model for a SC consisting of one manufacturer and multiple suppliers under a demand uncertainty scenario. To obtain an equilibrium condition, they proposed a quantity discount scheme embedded into the leader's (manufacturer's) function. Recently, Golpîra and Javanmardan (2021) proposed a novel risk-based robust modeling approach for a decentralized closed-loop SC system considering bi-level multi-objective programming under demand uncertainty. They compared the model results with deterministic decentralized structure and with centralized structures.

However, most approaches for DSC modeling using a BLPP approach, faced difficulties in making leader–follower decisions among their members. Caldentey and Wein (2003) showed that, in the case of a Stackelberg policy, a supplier's leadership power caused more harm to the SC and increased total system cost than a retailer's. On the other hand, customer service was hampered when the retailer was the leader, as less inventory than the required base-stock level was held. A similar result was found by Jemaï and Karaesmen (2007). Applying a Stackelberg game policy showed that if one party was more powerful in leading the inventory decisions, the inefficiency of a decentralized operation could be significant. A small number of researchers dealt with three or four stages SCs by using a bi-level approach, which again used an iterative approach to define a leader or follower (i.e., the follower of one step became the leader in the next step).

(b) **Nash Game Approach**

Some researchers applied a Nash game approach to finding solutions between two or more competing parties in a DSC. Game theory is a solution approach for a non-cooperative game involving two or more players, in which each player is assumed to know the strategies of the other players (Yue and You 2014). The NE can be defined as the stable state of a game in which a player cannot improve its payoff unilaterally, given that the actions of its rivals remain unchanged.

A Nash game model has also been applied under uncertain conditions in DSC planning. Gurnani and Gerchak (2007) developed a coordination model for a decentralized assembly system under an uncertain random yield. They proposed a Nash game nested within a Stackelberg game while considering the assembly firm as a leader and multiple suppliers as followers. As the suppliers made their decisions simultaneously and their costs were interrelated, they developed an optimal Stackelberg ordering policy for the assembly firm under the suppliers' NE condition. Wu et al. (2009) studied the NE of an industry with two competing SCs consisting of manufacturer-retailer operating in the same market under an uncertain demand scenario. They considered three SC strategies: vertical integration, manufacturer's Stackelberg, and bargaining on the wholesale price. Leng and Parlar (2010) developed two game-theoretic mathematical models for a multiple-supplier, single manufacturer assembly SC system that considered price-sensitive random demand. One was a leader–follower game that assumed the manufacturer was the leader and suppliers the followers, while the other was a NE game in which a manufacturer and suppliers simultaneously made their decisions without any form of communication. They proposed buyback and lost-sales cost-sharing contracts to ensure coordination among the members. Zhao et al. (2013) considered a closed-loop DSC with one manufacturer and two different competitive retailers under various uncertain parameters. They used both game theory and fuzzy theory to determine optimal pricing and the remanufacturing decisions of each member of the problem. They analyzed three cases: manufacturer Steckelberg, retailer Steckelberg, and vertical Nash (every firm has equal bargaining power and makes decisions simultaneously). They conducted numerical analyzes and summarized the findings under different scenarios. Chen et al. (2016) studied a SC with a single manufacturer and a single retailer under three different decentralized power structure scenarios under uncertainties for high-tech short life cycle products. They studied a manufacturer- Stackelberg (the manufacturer holds more bargaining power), retailer-Stackelberg (the retailer holds more bargaining power) and vertical Nash game in which each member had equal bargaining power who made their decisions simultaneously. They considered consumer demand, manufacturing cost, and sales effort cost as uncertain variables and obtained equilibrium decisions for each scenario. Comparative analyzes were conducted to analyze the effects of uncertainties on different power structured SCs. A few researchers investigated uncertainties that could occur in every stage of a SC. For example, Giri and Bardhan (2017) studied a three-echelon SC with uncertainty at every stage. They analyzed two different power structures of a decentralized channel, namely a Stackelberg game strategy and a vertical Nash strategy, assuming the upstream member as the Stackelberg leader between any two adjacent entities. Later, Gupta et al. (2018) developed Stackelberg game policies while assuming each member of a serial DSC as a leader and a vertical Nash game in which each member had equal power. They considered the uncertainty associated with customer demand, marginal production costs, and effort costs as linguistic or fuzzy variables. According to the model, under the Nash game setting, the supplier chooses its unit price, the manufacturer chooses its profit margin and quality efforts, and the retailer simultaneously chooses its profit margin and advertising efforts. They showed that both the quality and advertising efforts, as well as the total SC profit, were highest for the vertical Nash bargaining game for various DSC cases. In recent years, Modak and Kelle (2019) studied the effect of stochastic customer demand on a dual channel two-stage (manufacturer-retailer) DSC while considering both Stackelberg game (manufacturer leader) and Nash game settings. They proposed a hybrid all-unit quantity discount, along with a franchise fee contract to achieve chain coordination. Their models considered both known and unknown distributions of the random variables.

3.1.4 Summary of Literature Review of Quantitative Models

Based on the models adopted, a summary of the literature review for DSC planning is presented in Table 1. From the table, it can be seen that several studies considered SC coordination through various types of contract negotiations or coordination through the hierarchy of a chain. This type of modeling was limited to mostly two-stage SC structures which were vertically integrated. Moreover, these approaches required full information-sharing among the interacting partners, which is a major restriction for

Table 1 Summary of DSC planning models considering different modeling approaches

Model/strategy	Description	References to literature	Remarks
Contract negotiation	A formal agreement between SC partners to determine decision parameters to ensure total chain coordination	Lee and Whang (1999), Chen (2003), Krishnan et al. (2004), Giannoccaro and Pontrandolfo (2004), Gerchak and Wang (2004), Zhou et al. (2008), Hou et al. (2009), Proch et al. (2017), Jung et al. (2008), Bernstein and Federgruen (2005), He and Zhao (2012), Hu et al. (2013), Taghipour and Frayret (2013), Asian and Nie (2014) Heydari and Norouzinasab (2015), Fu et al. (2015), Muzaffar et al. (2017), Arikan and Silbermayr (2017), Basu et al. (2018), Giri and Sarker (2019)	Difficult to establish a contract among multiple interacting partners in a multi-stage SC
Hierarchical coordination	Finding optimized decisions for each stage separately and then coordinating them by exchanging output or establishing some other strategies according to a hierarchical structure	Lee and Whang (1999), Nagurney et al. (2002), Selim et al. (2008), Geng et al. (2010), Francis Leung (2010), Baboli et al. (2011), Larbi et al. (2012), Duan and Warren Liao (2013), Taghipour and Frayret (2013), Gansterer and Hartl (2020), Fu and Ma (2019), Jalil et al. (2018)	Limited to vertically integrated SC structures
Bi-level programming approach	The DM at the upper level (leader) first specifies a strategy, and then the DM at the lower-level (follower) specifies a strategy to optimize its objective with full knowledge of the action of the DM at the upper level	Cao and Chen (2006), Roghanian et al. (2007), Calvete et al. (2008), Yu et al. (2009a, b), Yu et al. (2009a, b), Hsieh and Lu (2010), Xiao et al. (2010), Naimi Sadigh et al. (2012), Ma and Wang (2013), Qiang et al. (2013), Tang et al. (2014), Yeh et al. (2014), Mokhlesian and Zegordi (2014), Yin et al. (2014), Zhu (2015), Schmitt et al. (2015), Shao et al. (2015), Taleizadeh et al. (2016), Schildbach and Morari (2016), Sarkar et al. (2016), Geunes et al. (2016), Hjaila et al. (2016),Yue and You (2017), Wang et al. (2017), Gao and You (2018), Gao and You (2019), Alamdar et al. (2019), Luo et al. (2019), Ezimadu (2020), Li et al. (2020), Glock et al. (2020), Nourifar et al. (2020), Golpîra and Javanmardan (2021)	Benefits one stage more than others and arise problems of making leaders/ followers among multi-stage SC members

(continued)

Table 1 (continued)

Model/strategy	Description	References to literature	Remarks
Combination of Stackelberg and Nash game	Combining Stackelberg game policy with Nash game to establish coordination among decentralized stages	Cachon (1999), Yang and Bialas (2007), Gurnani and Gerchak (2007), Wu et al. (2009), Leng and Parlar (2010), SeyedEsfahani et al. (2011), Ang et al. (2012), Gallego and Talebian (2013), Zhao et al. (2013), Mahdiraji et al. (2014), Taleizadeh and Noori-daryan (2014), Yue and You (2014), Chen et al. (2016), Chen and Xiao (2017), Giri and Bardhan (2017), Gupta et al. (2018), Modak and Kelle (2019), Mahmoodi (2020)	Requires knowledge of other players' strategies
Multi-objective optimization	Considering multiple objectives in each stage of a SC	Sabria and Beamon (2000), Roghanian et al. (2007), Raj and Lakshminarayanan (2008), Mahnam et al. (2009), Toksarı and Bilim (2015), Alaei and Setak (2015), Ben Abdelaziz and Mejri (2016), Soleimani et al. (2017), Nematollahi et al. (2018)	Very few research studies were found in the literature

decentralized structures. A significant amount of research on modeling decentralized structures using game theories, especially with BLPP or Stackelberg leader–follower strategies assuming one particular channel member(s) as a leader(s) and the other(s) as a follower(s) was found. However, selecting a leader/follower among multiple stages was found as one of the major challenges in using this strategy for a SC with equally powerful members. In contrast, a few studies were found that combined the Stackelberg and the Nash game approaches, considering the competition among the interacting members in a chain with uncertain conditions. However, using Nash game strategies requires knowing the possible reactions of the other partners' (/players'), which creates difficulties in modeling real DSC scenarios. Therefore, it can be said that the literature on appropriate quantitative modeling for multi-stage DSCs while considering restricted information flows and the equally powerful natures of the members under various uncertain scenarios, is limited.

3.2 Solution Approaches

In the literature, many solution approaches have been applied to solve models with DSC planning problems while considering uncertain parameters. Researchers used traditional optimization techniques and different types of heuristics and metaheuristics as solution methodologies. Also, simulation and simulation-optimization are two major solution tools that have gained a great deal of interest for solving complex, large-scale SC problems in real-life scenarios. Therefore, the solution approaches adopted for this research domain can be classified as follows: exact approaches, heuristics, metaheuristics, simulation, and simulation-optimization approaches. Some important research that used these approaches are discussed in the following sub-sections.

3.2.1 Exact Approaches

In the literature, various exact approaches were used to solve DSC systems. Due to the nature of complexities associated with decentralized planning problems (Bitran and Yanasse 1982), there are very few studies in which the usual linear programming (Bialas and Karwan 1984) or integer programming (e.g., branch-and-bound, simplex, etc.) techniques were used. Among the existing studies that used BLPP or Stackelberg game on DSC modeling considering uncertainties, some used a backward induction procedure to find equilibrium decisions, in which, firstly, a follower's problem was solved, then a leader's problem was solved using the response function of the follower (Xiao et al. (2010), Fu et al. (2015), Chen et al. (2016), Giri and Bardhan (2017)). Besides, optimality conditions were also used to obtain the equilibrium condition of bi-level decentralized problems (Hsieh and Lu (2010)). However, the most common method for solving a bi-level program involves finding the optimal conditions of the lower-level through a Karush-Kuhn-Tucker (KKT) conversion, where the lower-level problem is replaced with a single-level one. For example, Ang et al. (2012) used both Lagrange multipliers and KKT conditions to solve a multi-leader single-follower SC game model to derive the optimal solution. Yeh et al. (2014) solved bi-level programs by finding the optimal conditions of the lower-level through KKT conversion. They introduced these KKT conditions into the top-level as constraints, thereby creating a single-level non-linear program (NLP). Gao and You (2019) applied KKT conditions and Glover's linearization method, to reformulate a mixed-integer non-linear bi-level problem into a single-level stochastic mixed-integer linear program.

3.2.2 Heuristics

Heuristics can be defined as a subset of strategies that may ignore some information to make decisions more quickly and/or accurately than other complex methods (Gigerenzer and Gaissmaier 2011). Several research works developed heuristics to solve SC mathematical models. Usually, heuristics are used to solve complex mathematical models that consist of large SC networks. For instance, Yugang et al. (2006) developed an algorithm for determining the equilibrium of a manufacturer-retailers Stackelberg game in a decentralized environment. Taghipour and Frayret (2013) proposed a dynamic mutual adjustment search heuristic for coordinating the operational plans on a rolling horizon basis between two independent SC partners (manufacturer-supplier), linked by material and non-strategic information flows. An iterative solution algorithm was proposed by Heydari and Norouzinasab (2015) to establish a coordination mechanism for a two-echelon SC. Apart from using heuristics to solve the models using the BLPP approach, researchers also proposed heuristic approaches for other types of quantitative models to coordinate the independent entities of a decentralized network. Nematollahi et al. (2017) proposed a search procedure to coordinate a supplier-retailer decentralized pharmaceutical SC. Jokar and Hosseini-Motlagh (2019) proposed a heuristic search procedure for determining a coordination mechanism combining the buyback and wholesale price discount contracts within a manufacturer-retailer decentralized structure.

3.2.3 Metaheuristics

Metaheuristics are solution methods that interact between local procedures and higher-level strategies to create a process capable of obtaining robust solutions (Gendreau and Potvin 2010). Although heuristics and metaheuristics approaches do not guarantee to find a global solution, they are used to solve large-scale complex problems, which would be difficult to solve by other approaches. In particular, as many studies used BLLP approaches to solve DSC problems, metaheuristics were widely employed to solve such complex non-linear problems that exact approaches could not solve. However, fewer studies are found using metaheuristics approaches under uncertain conditions compared to deterministic ones. For example, Geng et al. (2010) used approximate dynamic programming procedures to obtain the system performances of a single distributor multi-retailer decentralized inventory system. Under uncertainties, researchers significantly used fuzzy programming to obtain feasible solutions for quantitative models. Roghanian et al. (2007) converted their developed bi-level problem into an equivalent deterministic model at each level and used fuzzy programming techniques to solve a multi-objective non-linear programming problem to obtain feasible solutions. Toksari and Bilim (2015) proposed a fuzzy goal programming approach to achieve the highest degree of each membership goal for a SC with multiple DMs.

3.2.4 Simulation

In the literature, simulations were used to imitate real-world scenarios. Simulations are computer-based tools that help DMs to analyze and improve system efficiency under uncertainties to represent practical scenarios (Keskin et al. 2010). Studies are using various simulation techniques to model decentralized systems. Rao et al. (2003) developed a simulation-based approach for a non-cooperative decentralized inventory-planning problem consisting of N-retailers and W-warehouses. They used Reinforcement Learning, which is a simulation-based stochastic optimization approach for finding near-optimal solutions. Fattahi et al. (2015) applied a Monte Carlo simulation to optimize a two-stage DSC under uncertainties. To deal with demand uncertainty, they generated scenarios using the Latin Hypercube Sampling method (Olsson et al. 2003), in which the number of scenarios was reduced using a forward reduction technique. Ben Abdelaziz and Mejri (2016) used a simulation-based solution algorithm for a decentralized non-linear multi-objective BLPP. Their approach showed a significant cost reduction in comparison to the usual inventory model.

3.2.5 Simulation-Optimization Approach

Although simulations have been widely used to imitate complex real-world scenarios, they may require large amounts of computational time, which causes difficulties in solving optimization problems. However, they can still be solved by simulation-optimization techniques (Keskin et al. 2010). Simulation-optimization is a method in which an optimization module is coupled with a simulation model (Wan et al. 2005). Many researchers found this to be a useful solution approach to solving practical optimization problems with real-world uncertain scenarios. For example, Acar et al. (2009) used an iterative solution approach that combined optimization and simulation methodologies to develop a decision support framework for a global SC network. Mahnam et al. (2009) made a hybridization of multi-objective particle swarm optimization (PSO) and simulation-optimization methods to solve their stochastic model. Similarly, Sahay and Ierapetritou (2014) proposed a hybrid simulation-based optimization framework for SC operational problems in an agent-based decentralized distributed network. They developed optimization and simulation models separately and then coupled them together in a hybrid approach to extract the benefits of both models.

Later, Shao et al. (2015) proposed a hybrid GA-based stochastic simulation approach to solve their model with an unlimited node expansion strategy. Their analysis revealed that appropriate parameters should be set according to the specific data environment during the use of this algorithm. Fattahi et al. (2015) developed an optimization-embedded simulation model to solve a retailer-manufacturer DSC with uncertain demand. They used a Monte Carlo simulation in the optimization model's policies while using two metaheuristic approaches—evolutionary strategies and imperialist competitive algorithm. They generated scenarios using the Latin Hypercube Sampling method with a scenario reduction technique to deal with demand uncertainty. In recent studies, Nourifar et al. (2020) developed a simulation-based hierarchical interactive PSO algorithm to solve a stochastic DSC model. They considered a tri-level decision procedure of three hierarchical levels of decision-making, where the leader was at the top-level (distribution center), the follower at the mid-level (manufacturer), and the sub-follower at the bottom level (supplier). They used a Monte Carlo simulation approach to search for an optimal solution to handle the uncertain parameters of the proposed model.

3.2.6 Summary of Literature Review of Solution Approaches

A summary of the literature review of different solution approaches used in this chapter's research domain is presented in Table 2. Notably, numerous researchers used different types of exact approaches, especially for BLPP with two or three-stage SC modeling problems. Moreover, a significant number of studies focused on heuristics or metaheuristics algorithms to solve complex models. Furthermore, in recent studies, increasingly more simulation and optimization approaches were used to solve large-scale models under uncertain real-world scenarios.

Table 2 Summary of DSC planning models considering different solution approaches

Solution approach	Description	References to literature	Remarks
Exact approaches	KKT approach, backward induction, optimality conditions, algebraic solution	Xiao et al. (2010), Hsieh and Lu (2010), Ang et al. (2012), Yeh et al. (2014), Fu et al. (2015), Chen et al. (2016), Giri and Bardhan (2017), Gao and You (2019)	Difficult to apply to large complex problems
Heuristics	Strategies for finding near-optimal solutions	Yugang et al. (2006), Taghipour and Frayret (2013), Heydari and Norouzinasab (2015), Taleizadeh et al. (2016), Yue and You (2017), Wang et al. (2017), Nematollahi et al. (2017), Jokar and Hosseini-Motlagh (2019)	Simple, easy to solve, requires less computational effort
Metaheuristics	Genetic algorithm, dynamic programming, fuzzy etc.	Roghanian et al. (2007), Geng et al. (2010), Toksarı and Bilim (2015)	Requires higher computational time
Simulations	Repetitive process for creating real-world scenarios	Rao et al. (2003), Fattahi et al. (2015), Ben Abdelaziz and Mejri (2016)	Close to real-world scenarios
Simulation-optimization	Optimization-embedded simulation process	Acar et al. (2009), Mahnam et al. (2009), Sahay and Ierapetritou (2014), Shao et al. (2015), Fattahi et al. (2015), Nourifar et al. (2020)	Useful approach under uncertainties requires less computational time

3.3 Application in Real-Life Situations

Applying quantitative mathematical models to real-life case studies is a useful way of validating them. In the literature, most studies have considered hypothetical data analyzes due to difficulties in data acquisition. However, very few studies have used real-life case studies to judge their stochastic DSC models. For instance, Ali et al. (2018) studied the effect of a demand disruption on the pricing and service strategies of a DSC using the Stackelberg (Manufacturer leader) game-theoretical approach and applied their proposed approach to a toy SC in Bangladesh, which comprised one manufacturer and two retailers. Bagul and Mukherjee (2019) presented a real-life automotive industry case to determine the sourcing strategy for a multi-tier decentralized supply network under demand uncertainty. Also, Gao and You (2019) presented a case study of a Marcellus shale gas SC using a game theory-based stochastic model under uncertainty for a multi-stakeholder non-cooperative SC system.

4 Summary of Review and Research Gaps

After reviewing the relevant literature, the following research gaps can be highlighted:

i. It is noted that most approaches that explored DSC systems considering various uncertain conditions considered specific assumptions and problem environments and used the game theory or BLPP techniques. Most of them assumed that, in a chain, any one member is a leader that makes its decisions first, while the other(s) act as a follower(s) that determines its decisions afterward. However, this type of problem environment may not be appropriate for many real-world DSC scenarios, in which every member in a chain acts independently without being dominated by any other(s).
ii. A significant number of studies were found that used a solution-exchanging mechanism among a chain's entities according to its hierarchical/sequential structure while using the output from one stage as an input to another. However, obtaining optimal solutions through this approach might not be suitable for a DSC structure as the stages make their own decisions independently without relying on those of others.
iii. Most research on DSC modeling adopted two or three-stage SC networks, not many considered a multi-echelon one consisting of multiple entities in each echelon. Some studies that involved a multi-stage SC used iterative solution mechanisms, in which followers of two interacting stages were assumed as leaders for the next stage(s).
iv. Although, as found in the literature, different contracts have been used as coordination mechanisms among individual entities and applied mainly for two-stage SCs, they can be difficult to implement among multiple stakeholders in a DSC system. Also, these contracts of SC coordination mechanisms forced a chain to

act as if it was vertically integrated, which might not be a true representation of common decentralized structures.

v. From the review, it is observed that most of the studies considered uncertainties in end-customer demand, while very few of them considered price and/or lead time uncertainties. Hence, more studies are needed to consider uncertainties in various parameters realistically in every SC stage.

In brief, significant research gaps were noticed in the literature on DSC modeling under uncertain conditions, regarding real-life structures and circumstances.

5 Future Research Directions

The analyzes of the articles reveal significant aspects in which future research could be directed. A few of them are listed below:

i. Studies could be conducted to develop quantitative models with more realistic assumptions and real-life problem scenarios representing practical DSC structures.
ii. Multiple independent stages should be considered in modeling approaches that address multi-tier complex DSC structures with multiple independent entities.
iii. Efficient solution algorithms could be developed to solve the stochastic quantitative models. Further studies could be performed using simulation and optimization approaches to capture real-world uncertainties in this context.
iv. Studies focusing on uncertainties and/or risks associated with different planning parameters for every SC member, other than only end-customer demand and/or supply, could be performed to understand real-world uncertain situations.

6 Conclusions

In this chapter, a literature review of studies in DSC planning under uncertainties and risks was presented. Though a significant number of papers have been found in the literature with SC modeling, review papers focusing on only DSC structures considering uncertainty and/or risks are scarce. This chapter focused on various quantitative models with different solution approaches for DSC systems under various uncertainties and risks. In summary, most of the literature on DSCs decision-making, either for cooperative or non-cooperative cases, considered game theory based modeling, particularly using a leader–follower strategy, in which the follower was forced to cooperate with the leader. However, assuming a leader or follower in any SC stage does not always represent the actual scenario of a decentralized structure. In particular, SC members' independent planning and autonomous nature rarely follow an unequal power distribution, as is often assumed in the literature. Moreover, this assumption becomes more complex in the case of a multi-stage SC. Also, selecting leader(s) or follower(s) among members is a major planning issue with additional computational challenges for these BLPP approaches. In the literature, modeling has also been conducted using Nash game approaches, whereby each player of different SC levels acts simultaneously with complete information about the game. However, having complete information about another player may not be possible in a decentralized setting due to confidentiality and/or the difficulties of sharing information with others. Some of the studies used contract negotiation methods for non-cooperative SCs, which mainly coordinated decisions for only a SC's adjacent entity, rather than all stages, and hence didn't address the full multi-stage DSC picture. In contrast, hierarchical or upstream planning with a sequential flow of iterative activities, also used by researchers, may not be suitable for a network with independent entities focused on their individual strategies while having only local information. Therefore, more research should be conducted on modeling DSC structures considering practical aspects and planning approaches. In addition, most of the studies in the literature considered mainly two-stage serial SCs with manufacturer-distributor, manufacturer-retailer, or distributor-retailer network systems, while leaving the consideration of complex multi-tier network structures for future research.

However, this study should admit its limitations. For example, the study was conducted based on research articles available in Google Scholar and Scopus, written in English. Thus, additional sources would possibly include further findings.

In conclusion, the models found in the literature mainly considered full information-sharing among entities and/or unequal power among SC's members, which conflicts with practical decentralized network structures. This assumption of a mismatch in power also hinders the development of strategies for achieving individual objectives in this type of DSC setting. Consequently, further studies should be performed while considering practical DSC structures and assumptions. In addition, it is noticed that although demand uncertainty was studied by many researchers, very few studies included other uncertain parameters in their planning models. Hence, demand uncertainty for each independent member and uncertainties in lead time, price, cost parameters, capacity, etc., should be considered for studying DSC systems. Also, few studies have been found with simulation approaches, although they can be a method of capturing real-world uncertain scenarios. Therefore, our review pointed out that more research is needed to develop realistic quantitative models and efficient solution approaches for ensuring coordination among independent stages of multi-echelon complex DSC systems to represent real-world uncertain scenarios.

Acknowledgements This research is partly supported by the ARC, Australia Discovery Project (DP210102939) grant awarded to R. Sarker and D. Essam.

References

Acar Y, Kadipasaoglu S, Schipperijn P (2009) A decision support framework for global supply chain modelling: an assessment of the impact of demand, supply and lead-time uncertainties on performance. Int J Prod Res 48(11):3245–3268. https://doi.org/10.1080/00207540902791769

Alaei S, Setak M (2015) Multi objective coordination of a supply chain with routing and service level consideration. Int J Prod Econ 167:271–281. https://doi.org/10.1016/j.ijpe.2015.06.002

Alamdar SF, Rabbani M, Heydari J (2019) Optimal decision problem in a three-level closed-loop supply chain with risk-averse players under demand uncertainty. Uncertain Supply Chain Manag 351–368. https://doi.org/10.5267/j.uscm.2018.7.002

Ali SM, Rahman MH, Tumpa TJ, Moghul Rifat AA, Paul SK (2018) Examining price and service competition among retailers in a supply chain under potential demand disruption. J Retail Consum Serv 40:40–47. https://doi.org/10.1016/j.jretconser.2017.08.025

Ang J, Fukushima M, Meng F, Noda T, Sun J (2012) Establishing Nash equilibrium of the manufacturer–supplier game in supply chain management. J Global Optim 56(4):1297–1312. https://doi.org/10.1007/s10898-012-9894-3

Arikan E, Silbermayr L (2017) Risk pooling via unidirectional inventory transshipments in a decentralized supply chain. Int J Prod Res 56(17):5593–5610. https://doi.org/10.1080/00207543.2017.1394586

Asian S, Nie X (2014) Coordination in supply chains with uncertain demand and disruption risks: existence, analysis, and insights. IEEE Trans Syst Man Cybern: Syst 44(9):1139–1154. https://doi.org/10.1109/tsmc.2014.2313121

Baboli A, Fondrevelle J, Tavakkoli-Moghaddam R, Mehrabi A (2011) A replenishment policy based on joint optimization in a downstream pharmaceutical supply chain: centralized versus decentralized replenishment. Int J Adv Manuf Technol 57(1–4):367–378. https://doi.org/10.1007/s00170-011-3290-x

Basu P, Liu Q, Stallaert J (2018) Supply chain management using put option contracts with information asymmetry. Int J Prod Res 57(6):1772–1796. https://doi.org/10.1080/00207543.2018.1508900

Ben Abdelaziz F, Mejri S (2016) Multiobjective bi-level programming for shared inventory with emergency and backorders. Ann Oper Res 267(1–2):47–63. https://doi.org/10.1007/s10479-016-2324-z

Bernstein F, Federgruen A (2005) Decentralized supply chains with competing retailers under demand uncertainty. Manag Sci 51(1):18–29. https://doi.org/10.1287/mnsc.1040.0218

Bialas WF, Karwan MH (1984) Two-level linear programming. Manag Sci 30(8):1004–1020. https://doi.org/10.1287/mnsc.30.8.1004

Bitran GR, Yanasse HH (1982) Computational complexity of the capacitated lot size problem. Manag Sci 28(10):1174–1186. https://doi.org/10.1287/mnsc.28.10.1174

Cachon GP (1999) Competitive and cooperative inventory management in a two-echelon supply chain with lost sales. Fuqua School of Business, pp 1–34. http://repository.upenn.edu/oid_papers/5

Cachon GP (2003) Supply chain coordination with contracts. In: Supply chain management: design, coordination and operation, vol 11, pp 227–339. https://doi.org/10.1016/s0927-0507(03)11006-7

Caldentey R, Wein LM (2003) Analysis of a decentralized production-inventory system. Manuf Serv Oper Manag 5(1):1–17. https://doi.org/10.1287/msom.5.1.1.12764

Calvete HI, Galé C, Mateo PM (2008) A new approach for solving linear bilevel problems using genetic algorithms. Eur J Oper Res 188(1):14–28. https://doi.org/10.1016/j.ejor.2007.03.034

Cao D, Chen M (2006) Capacitated plant selection in a decentralized manufacturing environment: a bilevel optimization approach. Eur J Oper Res 169(1):97–110. https://doi.org/10.1016/j.ejor.2004.05.016

Chen F (2003) Information sharing and supply chain coordination. In: A. G. d. K. a. S. C. Graves (ed) Handbooks in OR & MS, vol 11, pp 341–421. Elsevier. https://doi.org/10.1016/s0927-0507(03)11007-9

Chen K, Xiao T (2017) Pricing and replenishment policies in a supply chain with competing retailers under different retail behaviors. Comput Ind Eng 103:145–157. https://doi.org/10.1016/j.cie.2016.11.018

Chen L, Peng J, Liu Z, Zhao R (2016) Pricing and effort decisions for a supply chain with uncertain information. Int J Prod Res 55(1):264–284. https://doi.org/10.1080/00207543.2016.1204475

Chowdhury P, Paul SK, Kaisar S, Moktadir MA (2021) COVID-19 pandemic related supply chain studies: a systematic review. Transp Res Part E: Logist Transp Rev 148:102271. https://doi.org/10.1016/j.tre.2021.102271

Duan Q, Warren Liao T (2013) Optimization of replenishment policies for decentralized and centralized capacitated supply chains under various demands. Int J Prod Econ 142(1):194–204. https://doi.org/10.1016/j.ijpe.2012.11.004

Ezimadu P (2020) Modelling cooperative advertising decisions in a manufacturer-distributor-retailer supply chain using game theory. Yugoslav J Oper Res 1–33. https://doi.org/10.2298/YJOR181115001E

Fattahi M, Mahootchi M, Moattar Husseini SM, Keyvanshokooh E, Alborzi F (2015) Investigating replenishment policies for centralised and decentralised supply chains using stochastic programming approach. Int J Prod Res 53(1):41–69. https://doi.org/10.1080/00207543.2014.922710

Francis Leung KN (2010) A generalized algebraic model for optimizing inventory decisions in a centralized or decentralized multi-stage multi-firm supply chain. Transp Res Part E: Logist Transp Rev 46(6):896–912. https://doi.org/10.1016/j.tre.2010.03.003

Fu H, Ma Y (2019) Optimization and coordination of decentralized supply chains with vertical cross-shareholding. Comput Ind Eng 132:23–35. https://doi.org/10.1016/j.cie.2019.04.009

Fu H, Ma Y, Ni D, Cai X (2015) Coordinating a decentralized hybrid push–pull assembly system with unreliable supply and uncertain demand. Ann Oper Res 257(1–2):537–557. https://doi.org/10.1007/s10479-015-1865-x

Gallego G, Talebian M (2013) Multi-supplier and single retailer contracts: profit splits under equilibrium. In: 22nd National Conference of the Australian society for operations research, Adelaide, Australia

Gansterer M, Hartl RF (2020) The collaborative multi-level lot-sizing problem with cost synergies. Int J Prod Res 58(2):332–349. https://doi.org/10.1080/00207543.2019.1584415

Gao J, You F (2018) A Game theory approach to design and optimization of decentralized supply chains under uncertainty. In: 13th international symposium on process systems engineering (PSE 2018), pp 1603–1608. https://doi.org/10.1016/b978-0-444-64241-7.50262-7

Gao J, You F (2019) A stochastic game theoretic framework for decentralized optimization of multi-stakeholder supply chains under uncertainty. Comput Chem Eng 122:31–46. https://doi.org/10.1016/j.compchemeng.2018.05.016

Gendreau M, Potvin J-Y (2010) Handbook of metaheuristics, 2nd edn. Springer. https://doi.org/10.1007/978-1-4419-1665-5

Geng W, Qiu M, Zhao X (2010) An inventory system with single distributor and multiple retailers: operating scenarios and performance comparison. Int J Prod Econ 128(1):434–444. https://doi.org/10.1016/j.ijpe.2010.08.002

Gerchak Y, Wang Y (2004) Revenue-sharing versus wholesale-price contracts in assembly systems with random demand. Prod Oper Manag 13(1):23–33. https://doi.org/10.1111/j.1937-5956.2004.tb00142.x

Geunes J, Romeijn HE, van den Heuvel W (2016) Improving the efficiency of decentralized supply chains with fixed ordering costs. Eur J Oper Res 252(3):815–828. https://doi.org/10.1016/j.ejor.2016.02.004

Giannoccaro I, Pontrandolfo P (2004) Supply chain coordination by revenue sharing contracts. Int J Prod Econ 89(2):131–139. https://doi.org/10.1016/s0925-5273(03)00047-1

Gigerenzer G, Gaissmaier W (2011) Heuristic decision making. Annu Rev Psychol 62:451–482. https://doi.org/10.1146/annurev-psych-120709-145346
Giri BC, Bardhan S (2017) Sub-supply chain coordination in a three-layer chain under demand uncertainty and random yield in production. Int J Prod Econ 191:66–73. https://doi.org/10.1016/j.ijpe.2017.04.012
Giri BC, Sarker BR (2019) Coordinating a multi-echelon supply chain under production disruption and price-sensitive stochastic demand. J Ind Manag Optim 15(4):1631–1651. https://doi.org/10.3934/jimo.2018115
Glock CH, Rekik Y, Ries JM (2020) A coordination mechanism for supply chains with capacity expansions and order-dependent lead times. Eur J Oper Res. https://doi.org/10.1016/j.ejor.2020.01.048
Golpîra H, Javanmardan A (2021) Decentralized decision system for closed-loop supply chain: a bi-level multi-objective risk-based robust optimization approach. Comput Chem Eng 154, Article 107472. https://doi.org/10.1016/j.compchemeng.2021.107472
Gupta R, Biswas I, Kumar S (2018) Pricing decisions for three-echelon supply chain with advertising and quality effort-dependent fuzzy demand. Int J Prod Res 57(9):2715–2731. https://doi.org/10.1080/00207543.2018.1547434
Gurnani H, Gerchak Y (2007) Coordination in decentralized assembly systems with uncertain component yields. Eur J Oper Res 176(3):1559–1576. https://doi.org/10.1016/j.ejor.2005.09.036
Hafezalkotob A, Makui A, Sadjadi SJ (2011) Strategic and tactical design of competing decentralized supply chain networks with risk-averse participants for markets with uncertain demand. Math Probl Eng 1–27. https://doi.org/10.1155/2011/325610
Haque M, Paul SK, Sarker R, Essam D (2020) Managing decentralized supply chain using bilevel with Nash game approach. J Clean Prod 266:121865. https://doi.org/10.1016/j.jclepro.2020.121865
Haque M, Paul SK, Sarker R, & Essam D (2021) A combined approach for modeling multi-echelon multiperiod decentralized supply chain. Ann Oper Res. https://doi.org/10.1007/s10479-021-04121-0
He Y, Zhao X (2012) Coordination in multi-echelon supply chain under supply and demand uncertainty. Int J Prod Econ 139(1):106–115. https://doi.org/10.1016/j.ijpe.2011.04.021
Heydari J, Norouzinasab Y (2015) A two-level discount model for coordinating a decentralized supply chain considering stochastic price-sensitive demand. J Ind Eng Int 11(4):531–542. https://doi.org/10.1007/s40092-015-0119-5
Hjaila K, Laínez-Aguirre JM, Puigjaner L, Espuña A (2016) Scenario-based dynamic negotiation for the coordination of multi-enterprise supply chains under uncertainty. Comput Chem Eng 91:445–470. https://doi.org/10.1016/j.compchemeng.2016.04.004
Hou J, Zeng AZ, Zhao L (2009) Achieving better coordination through revenue sharing and bargaining in a two-stage supply chain. Comput Ind Eng 57(1):383–394. https://doi.org/10.1016/j.cie.2008.12.004
Hsieh C-C, Lu Y-T (2010) Manufacturer's return policy in a two-stage supply chain with two risk-averse retailers and random demand. Eur J Oper Res 207(1):514–523. https://doi.org/10.1016/j.ejor.2010.04.026
Hu F, Lim C-C, Lu Z (2013) Coordination of supply chains with a flexible ordering policy under yield and demand uncertainty. Int J Prod Econ 146(2):686–693. https://doi.org/10.1016/j.ijpe.2013.08.024
Jalil SA, Javaid S, Muneeb SM (2018) A decentralized multi-level decision making model for solid transportation problem with uncertainty. Int J Syst Assur Eng Manag 9(5):1022–1033. https://doi.org/10.1007/s13198-018-0720-2
JemaÏ Z, Karaesmen F (2007) Decentralized inventory control in a two-stage capacitated supply chain. IIE Trans 39(5):501–512. https://doi.org/10.1080/07408170601180536

Jokar A, Hosseini-Motlagh S-M (2019) Simultaneous coordination of order quantity and corporate social responsibility in a two-Echelon supply chain: a combined contract approach. J Oper Res Soc 71(1):69–84. https://doi.org/10.1080/01605682.2018.1524349

Jung H, Frank Chen F, Jeong B (2008) Decentralized supply chain planning framework for third party logistics partnership. Comput Ind Eng 55(2):348–364. https://doi.org/10.1016/j.cie.2007.12.017

Keskin BB, Melouk SH, Meyer IL (2010) A simulation-optimization approach for integrated sourcing and inventory decisions. Comput Oper Res 37(9):1648–1661. https://doi.org/10.1016/j.cor.2009.12.012

Krishnan H, Kapuscinski R, Butz DA (2004) Coordinating contracts for decentralized supply chains with retailer promotional effort. Manag Sci 50(1):48–63. https://doi.org/10.1287/mnsc.1030.0154

Larbi EYAS, Bekrar A, Trentesaux D, Beldjilali B (2012) Multi-stage optimization in supply chain: an industrial case study.In: 9th international conference on modeling, Optim, France

Lawton TC, Michaels KP (2001) Advancing to the virtual value chain: learning from the Dell model. Ir J Manag 22(1):91–112

Lee H, Whang S (1999) Decentralized multi-echelon supply chains: incentives and information. Manag Sci 45(5):633–640. https://doi.org/10.1287/mnsc.45.5.633

Leng M, Parlar M (2010) Game-theoretic analyses of decentralized assembly supply chains: non-cooperative equilibria versus coordination with cost-sharing contracts. Euro J Oper Res 204(1):96–104. https://doi.org/10.1016/j.ejor.2009.10.011

Leng M, Parlar M (2016) Game theoretic applications in supply chain management: a review. INFOR: Inf Syst Oper Res 43(3):187–220. https://doi.org/10.1080/03155986.2005.11732725

Li J, Luo X, Wang Q, Zhou W (2020) Supply chain coordination through capacity reservation contract and quantity flexibility contract. Omega. https://doi.org/10.1016/j.omega.2020.102195

Li X, Wang Q (2007) Coordination mechanisms of supply chain systems. Eur J Oper Res 179(1):1–16. https://doi.org/10.1016/j.ejor.2006.06.023

Liu J, Xiao T, Tian C, Wang H (2018) Ordering and returns handling decisions and coordination in a supply chain with demand uncertainty. Int Trans Oper Res 27(2):1033–1057. https://doi.org/10.1111/itor.12542

Luo H, Liu L, Yang X (2019) Bi-level programming problem in the supply chain and its solution algorithm. Soft Comput 24(4):2703–2714. https://doi.org/10.1007/s00500-019-03930-7

Ma W, Wang M (2013) Particle swarm optimization-based algorithm for bilevel joint pricing and lot-sizing decisions in a supply chain. Appl Artif Intell 27(6):441–460. https://doi.org/10.1080/08839514.2013.805596

Mahdiraji HA, Govindan K, Zavadskas EK, Hajiagha SHR (2014) Coalition or decentralization: a game-theoretic analysis of a three-echelon supply chain network. J Bus Econ Manag 15(3):460–485. https://doi.org/10.3846/16111699.2014.926289

Mahmoodi A (2020) Stackelberg-Nash equilibrium of pricing and inventory decisions in duopoly supply chains using a nested evolutionary algorithm. Appl Soft Comput 86:105922. https://doi.org/10.1016/j.asoc.2019.105922

Mahnam M, Yadollahpour MR, Famil-Dardashti V, Hejazi SR (2009) Supply chain modeling in uncertain environment with bi-objective approach. Comput Ind Eng 56(4):1535–1544. https://doi.org/10.1016/j.cie.2008.09.038

Mentzer JT, DeWitt W, Keebler JS, Min S, Nix NW, Smith CD, Zacharia ZG (2001) Defining supplychain management. J Bus Logist 22(2):1–25. https://doi.org/10.1002/j.2158-1592.2001.tb00001.x

Modak NM, Kelle P (2019) Managing a dual-channel supply chain under price and delivery-time dependent stochastic demand. Eur J Oper Res 272(1):147–161. https://doi.org/10.1016/j.ejor.2018.05.067

Mokhlesian M, Zegordi SH (2014) Application of multidivisional bi-level programming to coordinate pricing and inventory decisions in a multiproduct competitive supply chain. Int J Adv Manuf Technol 71(9–12):1975–1989. https://doi.org/10.1007/s00170-013-5601-x

Mourtzis D, Doukas M, Psarommatis F (2012) A multi-criteria evaluation of centralized and decentralized production networks in a highly customer-driven environment. CIRP Ann 61(1):427–430. https://doi.org/10.1016/j.cirp.2012.03.035

Muzaffar A, Deng S, Malik MN (2017) Contracting mechanism with imperfect information in a two-level supply chain. Oper Res Int J 20(1):349–368. https://doi.org/10.1007/s12351-017-0327-4

Naimi Sadigh A, Mozafari M, Karimi B (2012) Manufacturer–retailer supply chain coordination: a bi-level programming approach. Adv Eng Softw 45(1):144–152. https://doi.org/10.1016/j.advengsoft.2011.09.008

Nagurney A, Dong J, Zhang D (2002) A supply chain network equilibrium model. Transp Res Part E: Logist Transp Rev 38(5):281–303. https://doi.org/10.1016/S1366-5545(01)00020-5

Nematollahi M, Hosseini-Motlagh S-M, Heydari J (2017) Economic and social collaborative decision-making on visit interval and service level in a two-echelon pharmaceutical supply chain. J Clean Prod 142:3956–3969. https://doi.org/10.1016/j.jclepro.2016.10.062

Nematollahi M, Hosseini-Motlagh S-M, Ignatius J, Goh M, Saghafi Nia M (2018) Coordinating a socially responsible pharmaceutical supply chain under periodic review replenishment policies. J Clean Prod 172:2876–2891. https://doi.org/10.1016/j.jclepro.2017.11.126

Nourifar R, Mahdavi I, Mahdavi-Amiri N, Paydar MM (2018) Optimizing decentralized production-distribution planning problem in a multi-period supply chain network under uncertainty. J Ind Eng Int 14(2):367–382. https://doi.org/10.1007/s40092-017-0229-3

Nourifar R, Mahdavi I, Mahdavi-Amiri N, Paydar MM (2020) Mathematical modelling of a decentralized multi-echelon supply chain network considering service level under uncertainty. Sci Iran 27(3):1634–1654. https://doi.org/10.24200/sci.2018.50733.1842

Olsson A, Sandberg G, Dahlblom O (2003) On Latin hypercube sampling for structural reliability analysis. Struct Saf 25:47–68. https://doi.org/10.1016/S0167-4730(02)00039-5

Omar IA, Debe M, Jayaraman R, Salah K, Omar M, Arshad J (2022) Blockchain-based supply chain traceability for COVID-19 personal protective equipment. Comput Ind Eng 167:107995. https://doi.org/10.1016/j.cie.2022.107995

Proch M, Worthmann K, Schlüchtermann J (2017) A negotiation-based algorithm to coordinate supplier development in decentralized supply chains. Eur J Oper Res 256(2):412–429. https://doi.org/10.1016/j.ejor.2016.06.029

Qiang Q, Ke K, Anderson T, Dong J (2013) The closed-loop supply chain network with competition, distribution channel investment, and uncertainties. Omega 41(2):186–194. https://doi.org/10.1016/j.omega.2011.08.011

Queiroz MM, Ivanov D, Dolgui A, Fosso Wamba S (2020) Impacts of epidemic outbreaks on supply chains: mapping a research agenda amid the COVID-19 pandemic through a structured literature review. Ann Oper Res. https://doi.org/10.1007/s10479-020-03685-7

Raj TS, Lakshminarayanan S (2008) Multiobjective optimization in multiechelon decentralized supply chains. Ind Eng Chem Res 47:6661–6671. https://doi.org/10.1021/ie800153z

Rao JJ, Ravulapati KK, Das TK (2003) A simulation-based approach to study stochastic inventory-planning games. Int J Syst Sci 34(12–13):717–730. https://doi.org/10.1080/00207720310001640755

Roghanian E, Sadjadi SJ, Aryanezhad MB (2007) A probabilistic bi-level linear multi-objective programming problem to supply chain planning. Appl Math Comput 188(1):786–800. https://doi.org/10.1016/j.amc.2006.10.032

Sabria EH, Beamon BM (2000) A multi-objective approach to simultaneous strategic and operational planning in supply chain design. Omega 28:581–598. https://doi.org/10.1016/S0305-0483(99)00080-8

Sahay N, Ierapetritou M (2014) Hybrid simulation based optimization framework for centralized and decentralized supply chains. Ind Eng Chem Res 53(10):3996–4007. https://doi.org/10.1021/ie403395p

Sarkar B, Saren S, Sarkar M, Seo Y (2016) A stackelberg game approach in an integrated inventory model with carbon-emission and setup cost reduction. Sustainability 8(12). https://doi.org/10.3390/su8121244

Schildbach G, Morari M (2016) Scenario-based model predictive control for multi-echelon supply chain management. Eur J Oper Res 252(2):540–549. https://doi.org/10.1016/j.ejor.2016.01.051

Schmitt AJ, Sun SA, Snyder LV, Shen Z-JM (2015) Centralization versus decentralization: risk pooling, risk diversification, and supply chain disruptions. Omega 52:201–212. https://doi.org/10.1016/j.omega.2014.06.002

Selim H, Araz C, Ozkarahan I (2008) Collaborative production–distribution planning in supply chain: a fuzzy goal programming approach. Transp Res Part E: Logist Transp Rev 44(3):396–419. https://doi.org/10.1016/j.tre.2006.11.001

SeyedEsfahani MM, Biazaran M, Gharakhani M (2011) A game theoretic approach to coordinate pricing and vertical co-op advertising in manufacturer–retailer supply chains. Eur J Oper Res 211(2):263–273. https://doi.org/10.1016/j.ejor.2010.11.014

Shao J, Noche B, Sun Y (2015) Optimization of integrated supply chain planning under multiple uncertainty. Springer

Soleimani H, Govindan K, Saghafi H, Jafari H (2017) Fuzzy multi-objective sustainable and green closed-loop supply chain network design. Comput Ind Eng 109:191–203. https://doi.org/10.1016/j.cie.2017.04.038

Sun XT, Chung SH, Chan FTS, Wang Z (2018) The impact of liner shipping unreliability on the production–distribution scheduling of a decentralized manufacturing system. Transp Res Part E: Logist Transp Rev 114:242–269. https://doi.org/10.1016/j.tre.2018.04.002

Swaminathan JM, Smith SF, Sadeh NM (1998) Modeling supply chain dynamics: a multi-agent approach. Decis Sci 29:607–632. https://doi.org/10.1111/j.1540-5915.1998.tb01356.x

Taghipour A, Frayret J-M (2013) Dynamic mutual adjustment search for supply chain operations planning co-ordination. Int J Prod Res 51(9):2715–2739. https://doi.org/10.1080/00207543.2012.737952

Taleizadeh AA, Noori-daryan M (2014) Pricing, manufacturing and inventory policies for raw material in a three-level supply chain. Int J Syst Sci 47(4):919–931. https://doi.org/10.1080/00207721.2014.909544

Taleizadeh AA, Noori-daryan M, Govindan K (2016) Pricing and ordering decisions of two competing supply chains with different composite policies: a Stackelberg game-theoretic approach. Int J Prod Res 54(9):2807–2836. https://doi.org/10.1080/00207543.2016.1154621

Tang SY, Gurnani H, Gupta D (2014) Managing disruptions in decentralized supply chains with endogenous supply process reliability. Prod Oper Manag 23(7):1198–1211. https://doi.org/10.1111/poms.12160

Toksari MD, Bilim Y (2015) Interactive fuzzy goal programming based on jacobian matrix to solve decentralized Bi-level multi-objective fractional programming problems. Int J Fuzzy Syst 17(4):499–508. https://doi.org/10.1007/s40815-015-0036-1

Wan X, Pekny JF, Reklaitis GV (2005) Simulation-based optimization with surrogate models—Application to supply chain management. Comput Chem Eng 29(6):1317–1328. https://doi.org/10.1016/j.compchemeng.2005.02.018

Wang M, Zhang R, Zhu X (2017) A bi-level programming approach to the decision problems in a vendor-buyer eco-friendly supply chain. Comput Ind Eng 105:299–312. https://doi.org/10.1016/j.cie.2017.01.008

Wu D, Baron O, Berman O (2009) Bargaining in competing supply chains with uncertainty. Eur J Oper Res 197(2):548–556. https://doi.org/10.1016/j.ejor.2008.06.032

Wu Y, Wang J, Li C, Su K (2018) Optimal supply chain structural choice under horizontal chain-to-chain competition. Sustainability 10(5):1330. https://doi.org/10.3390/su10051330

Xiao T, Jin J, Chen G, Shi J, Xie M (2010) Ordering, wholesale pricing and lead-time decisions in a three-stage supply chain under demand uncertainty. Comput Ind Eng 59(4):840–852. https://doi.org/10.1016/j.cie.2010.08.011

Yang MH, Bialas W (2007) Nash-stackelberg equilibrium solutions for linear multidivisional multilevel programming problems. University at Buffalo Technical Report, Buffalo, NY, USA

Yeh K, Realff MJ, Lee JH, Whittaker C (2014) Analysis and comparison of single period single level and bilevel programming representations of a pre-existing timberlands supply chain with a new biorefinery facility. Comput Chem Eng 68:242–254. https://doi.org/10.1016/j.compchemeng.2014.05.025

Yin S, Nishi T, Grossmann IE (2014) Optimal quantity discount coordination for supply chain optimization with one manufacturer and multiple suppliers under demand uncertainty. Int J Adv Manuf Technol 76(5–8):1173–1184. https://doi.org/10.1007/s00170-014-6298-1

Yu Y, Chu F, Chen H (2009a) A stackelberg game and its improvement in a VMI system with a manufacturing vendor. Eur J Oper Res 192(3):929–948. https://doi.org/10.1016/j.ejor.2007.10.016

Yu Y, Huang GQ, Liang L (2009b) Stackelberg game-theoretic model for optimizing advertising, pricing and inventory policies in vendor managed inventory (VMI) production supply chains. Comput Ind Eng 57(1):368–382. https://doi.org/10.1016/j.cie.2008.12.003

Yue D, You F (2014) Game-theoretic modeling and optimization of multi-echelon supply chain design and operation under Stackelberg game and market equilibrium. Comput Chem Eng 71:347–361. https://doi.org/10.1016/j.compchemeng.2014.08.010

Yue D, You F (2017) Stackelberg-game-based modeling and optimization for supply chain design and operations: a mixed integer bilevel programming framework. Comput Chem Eng 102:81–95. https://doi.org/10.1016/j.compchemeng.2016.07.026

Yugang Y, Liang L, Huang GQ (2006) Leader–follower game in vendor-managed inventory system with limited production capacity considering wholesale and retail prices. Int J Log Res Appl 9(4):335–350. https://doi.org/10.1080/13675560600836910

Zaman F, Elsayed SM, Ray T, Sarker RA (2017) Co-evolutionary approach for strategic bidding in competitive electricity markets. Appl Soft Comput 51:1–22. https://doi.org/10.1016/j.asoc.2016.11.049

Zhao J, Liu W, Wei J (2013) Pricing and remanufacturing decisions of a decentralized fuzzy supply chain. Discret Dyn Nat Soc 2013:1–10. https://doi.org/10.1155/2013/986704

Zhou Y-W, Min J, Goyal SK (2008) Supply-chain coordination under an inventory-level-dependent demand rate. Int J Prod Econ 113(2):518–527. https://doi.org/10.1016/j.ijpe.2007.10.024

Zhu SX (2015) Integration of capacity, pricing, and lead-time decisions in a decentralized supply chain. Int J Prod Econ 164:14–23. https://doi.org/10.1016/j.ijpe.2015.02.026

Supply Chain Deep Uncertainties and Risks: The 'New Normal'

Derek Friday, Suzanne Ryan, Steven Alexander Melnyk, and Damon Proulx

1 Introduction

Conditions of supply chain deep uncertainty create avenues for the spread of overlapping opinions, and the tendency for risk managers, speculators, media houses, and researchers alike to conflate radical uncertainty with predictable risks (Townsend et al. 2018). For example, the limited understanding and inability for current supply chain risk management models to predict the consequences of the prevailing supply chain deep uncertainty resulted in misinterpreting the COVID-19 pandemic as a Black Swan (an unknown-unknown). This should not have been the case given the world's experience in managing previous flu pandemics. For instance, the centre for disease control and prevention website indicates communities and economies have survived several pandemics including the 1918 Pandemic (H1N1 virus), the 1957–1958 pandemic (H2N2), the 1968 pandemic (H3N2), and the 2009 H1N1 pandemic (Centres for Disease Control and Prevention 2022). Therefore, the existence of knowledge on previous pandemics works contrary to the Black Swan theory assumption on rare events, which are considered hard to predict, and beyond the

D. Friday (✉)
Waikato Management School, The University of Waikato, Hamilton, New Zealand
e-mail: Derek.Friday@waikato.ac.nz

S. Ryan · D. Proulx
Newcastle Business School, College of Human and Social Futures, The University of Newcastle, Callaghan, Australia
e-mail: Suzanne.Ryan@newcastle.edu.au

D. Proulx
e-mail: Damon.Proulx@newcastle.edu.au

S. A. Melnyk
Department of Supply Chain Management, Broad College of Business, Michigan State University, East Lansing, USA
e-mail: Melnyk@broad.msu.edu

S. K. Paul et al. (eds.), *Supply Chain Risk and Disruption Management*, Flexible Systems Management, https://doi.org/10.1007/978-981-99-2629-9_3

realm of human expectations, and can only be rationalised by hindsight (Vacante et al. 2012). This misinterpretation of states of knowledge across communities, industry practitioners, and researchers underpins experienced violations and misinterpretation of COVID-19 protocols and threatens the ability for businesses and governments to identify, distinguish, and model value-enhancing risk management decisions under different conditions of supply chain uncertainty.

Currently, the global supply chain environment remains plagued by multiple sources of stress, risk, and disruptions. As a result, businesses continue to face repeated, and potentially unprecedented risks which present an unusual situation where an appropriate response across multiple risk fronts is required to address persistent COVID-19 effects, escalating geo-political tensions and related politicization of supply chains through disinformation, and spiralling global inflation impacting leading economies especially in Europe, the USA, China, and Russia. Specifically, the simultaneous exposure to these sources of risk has reversed previous resilience gains global energy and food security (e.g., the '360°' reversal of energy policies which send Europe back to relying on fossil fuels on the backdrop of the 2050 net zero greenhouse gas (GHG) emission targets). Inability to mitigate the effects from the above conditions highlights the extent businesses and governments depend on predictable global supply chains to provide critical services and achieve gross domestic product targets in both emerging and developed economies. According to Bhattarai (2022), mixed messages about the global economy from reserve banks, international organizations such as the International Monetary Fund, and the World Bank and consultancy firms, among others make it difficult for businesses to forecast and respond to the evolving uncertainties and risks. Throughout the evolution of risk, it has been argued that understanding supply chain risks cannot be separated from 'uncertainty' (Prakash et al. 2017). Yet, there are ambiguities in the literature pointing to a misconception of radical uncertainty and risk as being synonymous, and reference to risk as inherent in uncertainty or the other way around (Simangunsong et al. 2012). According to the uncertainty continuum (Vilko et al. 2014), supply chain uncertainty could be viewed as a distinct concept from supply chain risk, with each form of uncertainty requiring specific strategies.

Consequently, businesses that survived the COVID-19 pandemic disruption are doing everything to stay afloat. These businesses commonly hold off renovations, hire casual workers instead of hiring full-time employees, stock lower-priced goods, and switch arrangements with big retailers such as Target to sell directly to consumers to gain control over production and profits. Despite these mitigation strategies, the global supply chain landscape continues to elevate drivers of deep uncertainty and risk to a level that is incompressible for the best financial or political analysts. For instance, throughout the COVID-19 outbreak, businesses and governments faced shortages of supplies and logistical bottles in international transportation never experienced before in the modern business world. Furthermore, the geo-political tensions leading to Russia's invasion of Ukraine brought to bear Europe's deep uncertainty and dependence on Russian gas, and the limited local capacity to quickly switch the oil and gas supply chain infrastructure (Corbeau 2022). The resulting outcome is the concentration and dependence of global supply chains on China and Russia for manufacturing, skilled cheap labour, rare minerals, wheat, oil, and gas which has further compounded the deep uncertainty and risks surrounding global supply chain networks. Similarly, the Centre for Strategic and International Studies estimates the global cybercriminal enterprise will cost businesses across the globe up to $5.2 trillion by 2024 (King and Gallagher 2020; Melnyk et al. 2021). Consequently, businesses continue to face losses worth billions of dollars due to malicious cyber activity threatening global supply chain ecosystems, with deep uncertainty and risk implications for communities, businesses, and governments.

In light of these significant issues, it is necessary to understand, where, if any, broad drivers of deep uncertainty and risk in the literature undermine the current scope of supply chain risk management models to appropriately mitigate against these factors. Therefore, this chapter aims to expand the application of supply chain risk management models and processes by examining the broad drivers of deep uncertainty and risk in the literature to determine where these issues exist, and, suggest potential solutions for future applications. Towards these ends, the chapter adopts the Web of Science (WoS) database as a source of academic supply chain uncertainty articles for analysis. The WoS database offers a broad coverage of academic journals and enables the categorisation of retrieved documents based on citations and publication trends (see Fig. 1). A search query was generated to search for articles on supply chain uncertainty in WoS: '*supply chain uncertainty*' *(Topic, Title, Abstract, Keywords).* The search for relevant articles on 11/02/2022 resulted in 636 documents categorised by WoS under seven document types: articles, proceeding papers, review articles, early access, corrections, editorial materials, and news items. Of the retrieved documents, 20 articles were excluded because they were either inaccessible or not published in English. As highlighted and discussed in the proceeding sections of this chapter, key findings include a significant increase in attention on supply chain uncertainty in the last five years. Additionally, there is a need to advance a larger portion of future studies towards focusing beyond finance and statistical forecasting, prediction, and optimisation techniques to address supply chain deep uncertainties.

2 Supply Chain Uncertainty Literature

The Publication and citation results from the retrieved articles for analysis indicate research attention on uncertainty and related risks remained low throughout the years until 2011 (see Fig. 1). The sharp increase from 2017 to 2021 can be attributed to a period when global supply chains for the first time were stressed by a simultaneous occurrence of multiple mega stressors such as COVID-19, climate change, heightened geo-political tensions and disinformation, broad cyberattacks and an urgent need to address these factors. The first set of articles addressing supply chain uncertainty begins with establishing empirical metrics for the measurement of uncertainty in customer satisfaction, demand, and inventory management. Firstly, Escudero et al. (1999) develop an empirical modelling framework for optimising product demand and lead times distribution to customers in the automotive industry. Applequist et al.

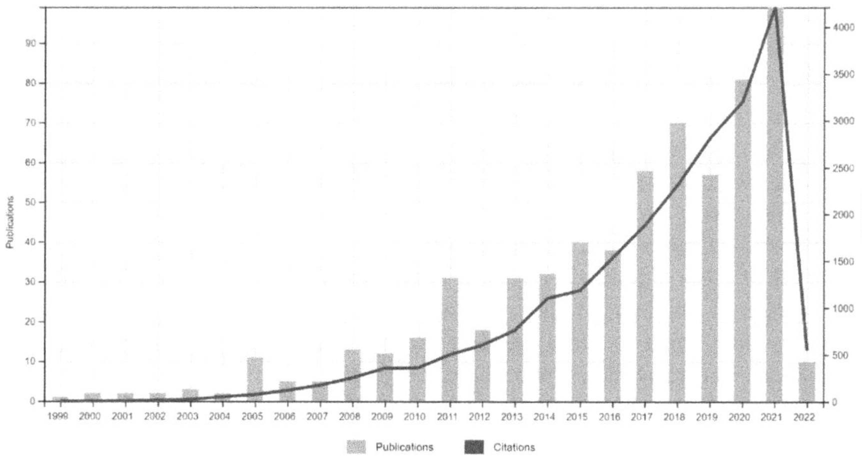

Fig. 1 Publication and citation trend of supply chain uncertainty articles retrieved from the WoS (1999–Feb 2022)

(2000) present a risk premium construct for estimating expected values and revenue attained in the event of demand uncertainties and associated issues with inventory variability within the chemical manufacturing industry. Gupta et al. (2000) adopt a framework of supply planning under demand uncertainty to forecast optimised levels of customer satisfaction.

While these studies represent the foundational contributions in the field of supply chain uncertainty, the most impactful contributions reside in the areas of developing decision frameworks for optimal technology choices in supply chain strategies across different firms. Proceeding research makes contributions towards advancing the development of machine learning algorithms for forecasting different environmental uncertainty problem parameters, as well as theoretical models of cost management within internal and external environmental constraints to the production and cost process in supply chain integration (Wong et al. 2011). The most recent trends in the field of supply chain uncertainty focus on understanding individual rationality in supplier, manufacturing and retailer decisions in operational decisions and contract structuring. Whereas other contributions focus on optimising profit maximisation through weighting supply chain coordination strategies and risks as these relate to financial instability (Li and Li 2022). Conversely, other contributions add to the application of optimisation principles to risk associated with supply sustainability across resource intermediaries, financial, social, logistical, and natural environmental factors (Raian et al. 2022).

Although Fig. 1 highlights a significant increase in attention on supply chain uncertainty in the last five years, the challenge across a larger portion of existing literature is the focus on advancing statistical forecasting, prediction, and optimisation techniques to address supply chain uncertainty. The advancement of big data analytic capabilities on which the current techniques are anchored could not save supply chains from global disruptions, especially in critical sectors such as humanitarian and medical supply chains pushed to a tipping point by the COVID-19 pandemic. In unique and unprecedented contexts such as that presented by COVID-19, existing techniques are adequate up to the point where existing data and experience can no longer be relied upon to generate probability distributions to inform decision making (Friday et al. 2021). Therefore, we argue that the inability of supply chain experts to address medical stocks, delayed deliveries, and port congestions, even with all the sophisticated I.T systems in place may be attribute to idiosyncratic responses that do not put into consideration negative supply chain externalities. For instance, on the backdrop of the COVID-19 pandemic, succeeding stressor such as the Russia-Ukraine war and escalating geo-political tensions between USA and Chain over the Taiwan Striates, revealed deeper supply chain uncertainties and interdependences that could never have been predicted or forecasted by strategic planning or existing algorithms. As such, it makes sense for both academics, practitioners, international organisations and regulators to review the techniques currently relied upon to manage supply chain uncertainty and related disruptions.

3 Levels of Supply Chain Uncertainty

A content analysis of the literature revealed varying positions around the definition of supply chain uncertainty. From the existing definitions and contexts, uncertainty can refer to the lack of knowledge about an event that reduces confidence in decision making and drawing conclusions from available data/information (Phillips et al. 2006). Whereas, 'supply chain uncertainty' is a situation in which the risk manager does not know definitely what to decide as they are indistinct about the objectives. This may include a lack of information about the supply chain environment, where managers are unable to predict the risk impact or possible control actions on supply chain operations behaviour (van der Vorst and Beulens 2002; Vilko et al. 2014). Supply chain uncertainty may also entail different classifications which can encompass endogenous and exogenous uncertainty (Trkman and McCormack 2009). In addition to supply and demand uncertainty (Gong et al. 2014), other forms of categorisation can include internal organisational uncertainty (emerges from the focal firm); internal supply chain uncertainty (emerges from within the control of the focal firm or its supply chain partners); and external uncertainties (emerging from outside the supply chain) (Simangunsong et al. 2012).

Two key studies share related views on understanding uncertainty based on a state of knowledge continuum. While Townsend et al. (2018) explain uncertainty based on ignorance/unknowingness such as ambiguity, equivocality, and complexity, a prior study by Vilko et al. (2014) offers a more detailed breakdown of supply chain uncertainty on a continuum ranging from complete certainty to radical uncertainty. Results from the content analysis highlight the different levels of supply chain uncertainty addressed over a 20-year (2001–2021) period (see Fig. 2). The results were attained by coding, querying the literature, and analysing six levels of the supply chain continuum respectively: from complete certainty, probabilistic

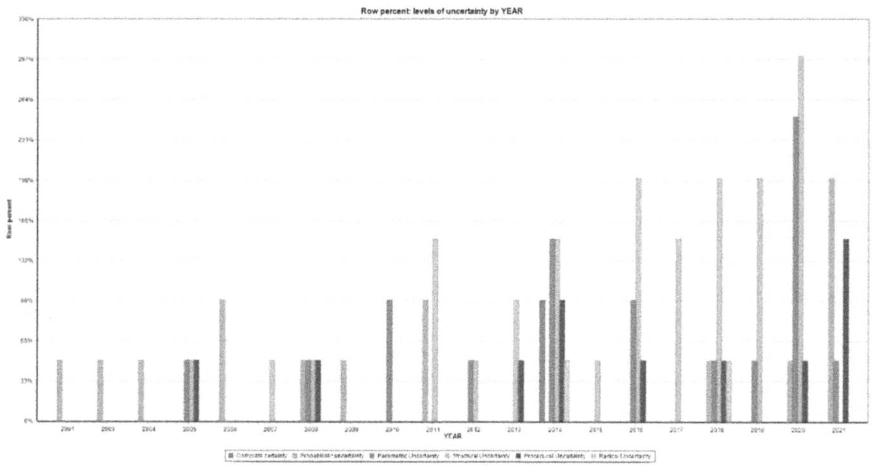

Fig. 2 Representation of different levels of the supply chain uncertainty continuum in the literature (2001–February 2021)

uncertainty, parametric uncertainty, structural uncertainty, procedural uncertainty, to radical uncertainty.

The results indicate that for the most part of a decade (2001–2011), studies on supply chain uncertainty focused mainly on probabilistic uncertainty, where *Probabilistic uncertainty* refers to when alternative outcomes of decisions are known in terms of probabilities regarding the likelihood of achieving the outcome and impact. The analysis indicates that definitions of structural uncertainty (that are more complicated to understand) can be traced back to as early as 1984, yet the concept is not picked up in supply chain uncertainty until 2005. *Structural uncertainty* refers to the knowledge related to the state of the world in a future that is imperfect, and only subjective beliefs can be projected. At this level of uncertainty, analyses cannot objectively assess the probability of alternative choices or their causality.

Although the two extremes of the supply chain uncertainty continuum (complexity and radical uncertainty) are contemporary terms, they have attracted limited research attention in a time when understanding how to make decisions in certain and extremely uncertain environments is critical. Conversely, *complete certainty* refers to an ideal hypothetical world in which the decision maker knows all relevant information; there are no actual risks, as each is rectified. On the other hand (Vilko et al. 2014), *radical uncertainty* refers to a hypothetical decision making world in which there is total imperfection in terms of knowledge about the structure or probability of alternatives (an example of the Black Swan theory). In the context of our analysis, the results indicate an urgent need for a shift from a 'certain' and 'probabilistic' understanding of supply chain uncertainty towards considering more radical levels of deep uncertainty in line with the current times: parametric, structural, procedural, and radical supply chain uncertainty earlier discussed. For example in the case of the COVID-19 pandemic or escalating geo-political tensions and related disinformation influence operations, managerial risk management decisions were undermined by limited understanding of parametric and procedural uncertainty. *Parametric uncertainty* refers to when the alternative outcomes of decisions/choices are known but the probability parameters are not: environment dependent. *Procedural uncertainty* refers to a situation where the supply chain risk manager is constrained by his or her computational and cognitive capabilities and therefore cannot form a clear picture of the decision making processes or the risk events, mainly on account of their complexity. As such, our proposal for a shift can be attributed to the growing need to pay attention to the more complex and overlapping drivers supply chain deep uncertainty and risks such as climate change, broad supply chain cyberattacks, war posturing and international wars, that do not conform to the interpretation of common supply chain risks identified from probability distributions, optimisation, and predication/forecasting techniques.

4 Multiple and Simultaneous Drivers of Deep Supply Chain Uncertainty and Risk

The troubling challenge in managing systematic supply chain disruptions in this era of digital and data-driven supply chains can be attributed to our inability to cope with global supply chain deep uncertainties. According to Cox (2012), the relevance of big data in predicting future outcomes is in doubt, as formerly relied upon and validated risk management models giving the probabilities of future performance are not readily available. There are conflicting views among experts on the probable consequences of alternative managerial decisions and government or international institutional policies. For example, unwarranted demands to reach consensus among decision bodies such as the UN security council, effects of groupthink biases among policymakers, and ambiguity around what decision models to use among risk analysts (e.g., central banks, security stock markets) have continuously undermined the ability to address overlapping drivers of deep uncertainty such the spiraling global inflation and effects from the Russia-Ukraine war. Similarly, the feelings of morale obligations not to oversimplify the analysis by imposing one specific solution continue to undermine responses to drivers of global supply chain deep uncertainties. With emotions and 'politics of the supply chain' running high, coupled with convictions of being right, simple questions such as where, when, and how to pre-empt drivers of supply chain deep uncertainty become more complex when multiple and simultaneous drivers of supply chain uncertainty simultaneously occur. This is apparent during difficult managerial decision making situations exacebated by climate change, pandemics, deliberately spread pathogens, state sponsored cyberattacks, or terrorist threats, in conjuction with trading partners advancing local and international core strategic interests, and protecting territorial integrity of sovereign states.

Therefore, the multiple, and simultaneous occurrence of mega supply chain risk and stressors is not only unprecedented, but also offers researchers and practitioners an opportunity to re-examine the efficacy of existing uncertainty and risk management techniques. Supply chain risk management approaches covering various traditional categories of risks (e.g., supply, demand, operations, environmental, legal, and technology) are well advanced (e.g., Bailey et al. 2019; Manuj and Mentzer 2008; Jüttner et al. 2003; Pournader et al. 2020). However, disruptions on the backdrop of the COVID-19 pandemic and escalatating geo-political tensions in the last 3 years are forcing supply chain risk managers and academics to re-examine how uncertainty and risk consequences due interconnectivity and interdependence in global supply chain operations can be managed. For instance, Fig. 3 highlights results from a content analysis of literature on supply chain uncertainty and shows how four simultaneously occurring drivers of supply chain deep uncertainty are represented in the literature: geo-political tensions, broad supply chain cyberattacks, natural calamities, and the COVID-19 pandemic. Prior to the Covid-19 pandemic, issues on climate change and cybersecurity dominated discussions in supply chain uncertainty. As will be discussed in the following section, we can confirm from looking at the graph in Fig. 3 that there hasn't been a time when all the four examined broad drivers of supply chain deep uncertainty are adequately represented in the literature (Fig. 4).

(a) *Pandemics* e.g., *COVID-19*

Several supply chain risk managers, industry experts, and governments a like were wrong in their interpretation of uncertainty surrounding the COVID-19 pandemic or simply reacted to the effects as they unfolded. Results highlighted in Fig. 3 indicate supply chain uncertainty research paid more attention to pandemics in 2021 and pandemic are becoming a dominant issue in 2022. Yet, as early as December

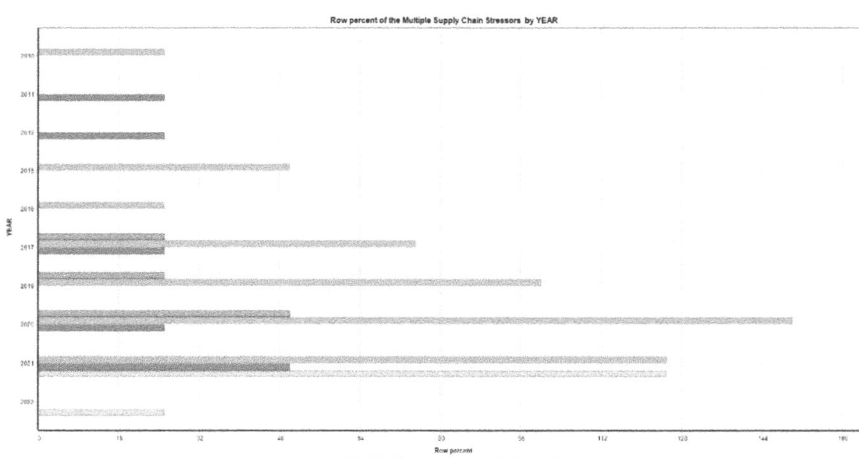

Fig. 3 Multiple and simultaneous drivers of deep supply chain uncertainty addressed in the literature over a period of 23 years (1999–February 2022)

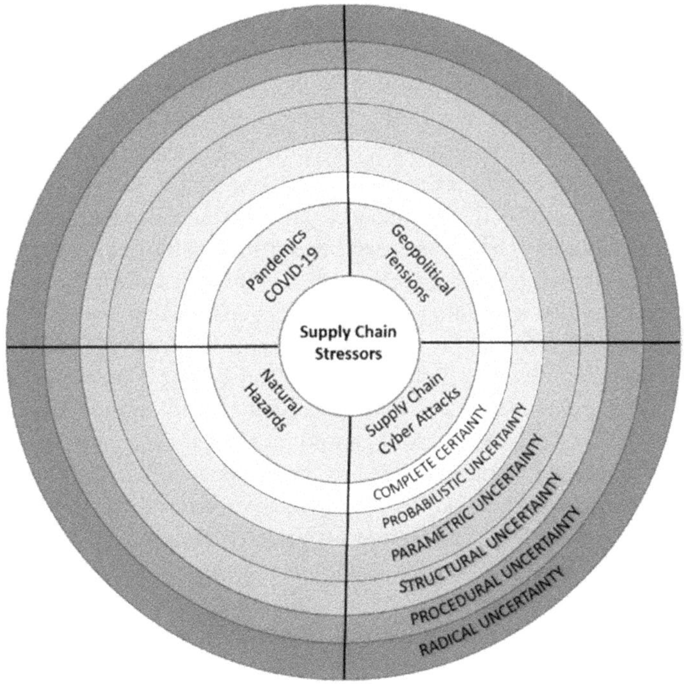

Fig. 4 Illustration of an association between mega supply chain stressors and different levels of uncertainty

2019 during the COVID-19 outbreak, businesses and governments faced shortages of medical and raw material supplies, logistical bottles in international transportation, and sporadic panic buying behaviour. In addition to huge losses of human lives and shortage of skilled labour, critical for supply chain risk managers was the realisation of the extent of concentration and deep dependence of local supply chains on China, especially for manufactured goods, and raw materials in the form of rare earth metals. Physical measures such as travel bans by the USA and the UK targeted communities from countries such as South Africa not only severely impacted tourism supply chain chains, but also resurrected nationalistic ideologies in local manufacturing that we thought were long gone due to global sourcing. The share of total GDP generated from travel and tourism globally fell from 10.3% in 2019 to 5.3% in 2020 following the pandemic outbreak and increased by just about 1–6.1% in 2021 (Statista 2022). Similarly, as of FY2022, countries such as Australia and USA and others in Europe were focusing on exploring local production capacity and the urgent need to localise supply chains to address the dependency on global supply chain networks. According to Friday et al. (2021), addressing deep uncertainty effects driven by the COVID-19 pandemic will require broadening steps in supply chain risk management processes to include a baseline capturing global supply chain stressors regardless of whether managers are responding to local drivers of uncertainty and risk.

(b) *Geo-political tensions*

Geo-political tensions have received low but consistent research attention from 2011 to 2020, and with a significant increase in 2021 (see Fig. 3). Geo-political tensions on the backdrop of the COVID-19 pandemic have soared since Russia's invasion of Ukraine. As global supply chain managers muddle through the likely effect of a potential conflict between the USA and China over the Taiwanese Striates, investors and policymakers worry that deepening tensions among leading nuclear power countries could exert a drag on global supply chains and economies already pushed to a tipping point by COVID-19, spiralling global inflation, and a sharp increase in uncertainty and risks with severe consequences for critical sectors such as the medical industry (Caldara et al. 2022). Several drivers of geo-political tension are apparent in existing research: petrochemical contexts, trade sanctions between USA and allies versus China and Russia on the other end, are among other examples spanning from foreign occupation, conflicts, to trade restrictions historically (Helbig et al. 2016).

While cross-jurisdictional tension, trade disagreements and international wars have been engrained in human function for millennia, today's global supply chains are not immune to war effects as the global economy is inherently interconnected and interdependent. Within this, both governments and businesses are exposed to several geo-political risks and uncertainties. According to Corbeau (2022), up to 36% of Europe's gas demand is supplied by Russia, with leading European economies such as Germany offering no immediate solution in sight to develop domestic capacity. For instance, in assessing geo-political tensions between the United States and China trends and risks, Wolf and Kalish (2021) highlight several factors that could inform how supply chain risk managers respond to deep supply chain deep uncertainties. In particular, the issue of restricting access to information and framing biases (disinformation) around issues such as USA's ambiguous policy and relationship with Taiwan versus the 'one China principle' which exacerbate the level of supply chain uncertainty that risk managers must consider. The following scenarios are likely to influence supply chain risk management strategies in the future:

- The decoupling scenario allowing businesses and countries to pursue policies that reduce dependence in the production of goods of national importance by countries considered unfriendly.
- Attempts to gain and preserve dominance in critical sectors by thwarting the other's efforts to access critical assets such as intellectual property and semiconductor supply chains.
- Dimishing reliance on third-party suppliers including heavy scrutiny of business ownership structures, and strengthening of data localization and privacy regulations.
- Data localization regulations will become more stringent as countries attempt to prevent non friendly countries from obtaining sensitive information, thus deepening global supply chain uncertainty.

Currently, the concentration of local and international parts of supply chains could act as a deterrence to war between the two largest economies of China and the USA. For example, China controls a substantial fraction of the world's mineable lithium supply and processes 80% of all minerals used in lithium-ion batteries, in addition to housing a big percentage of the world's manufacturing capacity. Following Europe's experiences from the Russian-Ukraine war, dependence and centralisation of supply chain operations in countries such as Russia and China explain why supply chain risk managers are grappling with the urgent need to decouple and localise operations in vain. In addition to restricting access to critical components (e.g., earth metals) and affordable oil and gas, any geographical shift in just one part or an entire supply chain to avoid impacts from geo-political tensions can create inconceivable knock-on effects and varying levels of deep uncertainty for a local manufacturing and supply chain ecosystem.

(c) *Broad supply chain cyberattacks*

Cybersecurity is a well established field in Information and Communication Technology (ICT), and is primarily addressed internally within the firm. It's only within the last decade that researchers and industry experts have recognised the urgent need to understand the breadth of negative externalities associated with cybersecurity across the supply chain. From the analysis results highlighted in Fig. 3, there is very little attention in literature (from 2017 to 2020) on cybersecurity as a driver of supply chain deep uncertainty and risk. Yet, several broad cyberattacks demonstrate how to focus on cybersecurity as a supply chain issue raises deep uncertainty and risks for targeted supply chains or customers who make use of vulnerable suppliers. For example, a cybersecurity investigation by FireEye in 2020 uncovered a broad and sophisticated state-sponsored attack, that not only uncovered the SolarWinds supply chain attack, but also a deeper indirect effect on government agencies and consumers reliant on SolarWinds as a major American information technology firm. Cybercriminals inserted malicious code in SolarWinds software to compromise over 33,000 companies that rely on SolarWinds to manage their IT resources. Until FireEye reported the breach to the U.S. National Security Agency (NSA), NSA had no clue of the threat or extent of the breach across impacted U.S. Federal agencies: Energy Department, the National Nuclear Security Administration, the Department of Déefense, the Pentagon, the Department of Justice, and the Department of Homeland Security (Melnyk et al. 2021). Other broad supply Chain cyberattacks such as NotPetya exemplify the deep uncertainty and potential impact caused by malware indirectly spreading along a global supply chain network. A report by European Network and Information Security Agency indicates that up to 66% of analysed incidents focused on supply chain digital assets (e.g., suppliers' code) in order to compromise a prime target.

What the above discussion suggests is that a manufacturer with limited ability to address cyber supply chain risks beyond the 1st tier supplier or customer faces significant uncertainty, mainly because of the uncertainty surrounding the high likelihood of an indirect cyberattack through weak points in the supply chain. The key concern for a focal firm in a multi-tier supply chain is that suppliers are at times unaware of indirect supply chain cyberattacks or may choose not to be transparent about how they were compromised (ENISA 2021). Critical among the cybersecurity concerns discussed above is that increasing deep supply chain uncertainties are a result of persistent indirect cyberattacks on targeted weak suppliers at the 3rd or even 4th tier in a multi-tier supply chain. Implying that the deep uncertainty and likelihood of successful indirect cyberattack through weak suppliers is no longer an issue of when but rather a 'new normal' that supply chain risk managers must deal with on a regular basis.

(d) *Climate Change—Natural Disasters*

The increasing focus on climate change as a driver of supply chain uncertainty since 2010 was highest in 2020 but is now at the verge of being overtaken by the focus on emerging geo-political tensions leading to the Russia-Ukraine war. Yet, uncertainty and risks resulting from global warming and impacting the supply chain financing and performance are critical for the sustainable growth of the global economy. The impact of climate change on the supply chain ecosystem are immense. For example, the World Wildlife Fund (WWF) for Nature-Australia and the University of Sydney estimated the 2019–20 bushfires to have cost Australian agriculture between $4 and $5 billion. The impact was estimated by examining damaged farm buildings and equipment, reduction in farmland values; loss of crops and livestock deaths; among others (Bishopa et al. 2021). According to the National Recovery and Resilience Agency, and the Australian Institute for Disaster Resilience web page, the impact of the summer bushfires on the New South Wales (NSW) ecosystem was unprecedented: 26 lives were lost, 2448 homes destroyed, and 5.5 Million of hectares of land were burnt, more than 3 Billion animals (including some rare or threatened plant, animal, and insect species, with the complete loss of some species believed to be permanent) are estimated to have been killed or displaced. The Insurance Council of Australia (ICA) declared the fires a catastrophe and enabled the processing of insurance claims from the bushfires in 2019–20 across NSW, Queensland, Victoria (VIC) and South Australia numbered 38,181, with estimated losses of $2.32 billion. In the first quarter of 2020, a Royal Commission into National Natural Disaster Arrangements was established, and in the same period, the Australian Government committed $2 billion to the National Bushfire Recovery Fund to assist impacted communities.

Similar to bushfire experiences in other countries such as the USA, risk management strategies aimed at mitigating the bushfire's impact on Australia's supply chain ecosystem are required. What is certain is that the 2019–2020 'Black Summer' bush fire season was the worst in NSW, with higher-than-average temperatures, low moisture levels following several years of drought fuelling the fires increasing in intensity, severity, and duration, driven in part by climate change. Besides wildfires, the Queensland 2011 floods highlighted significant weaknesses and need for strategies for supply chain security in the Australian food industry (McMahon et al. 2015). More recently, there has been an energy policy reversal back to fossil sources of energy (e.g., coal) among countries in Europe to address the deep uncertainty on the oil and gas supply from Russia. This presents a conundrum for risk managers aiming to address climate change challenges impacting supply chain operations, and the move to achieve the net zero GHG targets by 2050. Global warming has presented a situation where society encounters several unknown and increasingly naturally occurring risks (e.g., Fires, Floods and Rising Sea levels). However, insights presented by Europe's double standards in decision making around the dependence on Russian oil and gas highligt key supply chain risk management challenges, and reinforce the urgent need to keep track of the 2050 net zero GHG emissions. In addition to reverting fossil fuels, supply chain risk managers are unable to determine the extent politicians in Europe are willing to compromise the survial of manufacturing businesses or risk the lives of ordinary people through a cold winter due to lack of affordable gas, in order to prove the accuracy and effectiveness of the sanctions aimed at crumbling Russia's economy. The promised future for renewable energy is at an all-time low.

Some commentators argue that the high cost of gas and electricity will spur investors and European governments to invest more in wind, solar and nuclear energy projects in friendly countries such as Australia that are naturally endowed with resources critical for renewable energy. While there is no sign of refraining from war among proponents of a USA led NATO and Russia supporters, the increasing level of disinformation surrounding the war (framing biases) has split the world opinions on how peace can be achieved. On one hand, you have supporters of USA/NATO/Europe funding the Ukraine war versed those who view Russia as a superpower that is responding to an existential threat, just like any other superpower would have done to protect her sovereign and territorial integrity and strategic interests. However, the resulting disruptions in good, oil, and gas supply chains continue to impact businesses and the lives of ordinary people globally. Additionally, with resources (e.g., lithium, cobalt, and rare minerals) for renewable energy located mainly in Russia, and China, while the advancement in R&D, technologiy innovations and patents is led by Western economies, the deep uncertainty caused by escalating geo-political tensions, international wars and the impact previously relied upon business relationships between trading countries, will have to be resolved for global supply chains to regain previous levels of stability and net zero 2050 GHG emission targets in line with the Paris climate change agreement.

5 The Underlying Consequences—Spiralling Global Inflation

Inflationary pressures on the backdrop of COVID-19 economic stimulation packages have continued to impact the global world economy, raising the uncertainty and risk of stagflation. Exacerbated by the Russia-Ukraine war, and global consumer price inflation rose above central bank targets in leading global economies. Inflation is envisioned to remain elevated for a longer time and at higher levels than any of the banks, international organisations and experts have been able to predict or forecast. Continuous adjustments of inflation targets and raising bank interest rests have failed to curb the spiralling global inflation due to supply chain bottlenecks and edging commodity prices including oil, gas, and energy. The Russia-Ukraine war is leading to high global commodity prices, increasing food insecurity, exacerbating inflation and tighter financial vulnerability, and heightening international policy on uncertainty. Due to the resulting deep uncertainty and risks, economics and finance institutions have failed time and time again to accurately predict global growth or to forecast the level of anticipated economic slowdown in 2022. According to Guénette and Wheeler (2022) emerging market and developing economies' output was expected to decline from 6.6% in 2021 to 3.4% in 2022 due to negative externalities from the Russia-Ukraine war and the resulting commodity price volatility, global trade disruptions, and tighter financing conditions. For example, prior to the invasion, the world economies were focusing on reversing the severe losses and disruptions from the pandemic such as capital and skilled labour constraints to support the global economic recovery amid uneven and lingering supply bottlenecks. However, the increasing focus and sanctions on containing Russia's war in Ukaraine and the resulting hire levels of inflation in Europe have distracted the world from this agenda.

Finance and trade sanctions by the United States and the European Union (EU) resulted in a boomerang effect with severe consequences for businesses and communities in countries like Germany that heavily depended on Russian oil and gas. A summary of the sanctions and trade restrictions include: freezing Russia's foreign exchange reserves, restricted access to banking networks, blocking transactions with entities in sanctioning countries including a cut off from the SWIFT financial messaging system, banning of new investments in the Russian energy, export and import restrictions focused on "dual-use" technologies including semiconductors, goods and services related to aviation, aerospace and oil and gas production, and luxury goods, and closing the airspace to Russian flights (Guénette et al. 2022). The sanctions did not only fail to cripple Russia's economy, the assigned 'selective default' credit ratings, and the high probability of future sovereign debt default, have all failed to materialise or deter emerging economies in Asia, Africa, South America, and the Middle-East from trading with Russia. Furthermore, attempts by the U.S. government to block Russian debt service payments made from dollar accounts held with U.S. financial institutions did not signal the anticipated likelihood of a sovereign default in the financial market. The Russian economy that was tipped to enter a recession emerged stronger in comparison to United Kingdom and some European countries on the verge of collapsing because of unprecedented high inflation and surging costs of energy impacting the manufacturing industry. More importantly, for this chapter is that experts in central banks and institutions such as the World Bank continue to get their inflation predictions wrong partily because of the deep uncertainty created by Russia's war in Ukraine. For impacted global and local supply chains facing spiralling inflation from the war's negative externalities, there is no option but to continuously plan and adapt to a business environment with deep uncertainty and related inflationary risks.

6 Managerial Implications—Responding to Supply Chain Deep Uncertainty

Managing supply chain disruptions under conditions of deep uncertainty requires broadening the risk management discussion beyond logic-based strategies based on scenario planning, forecasting, and optimisation processes based on known 'states of nature' (probability and risk) to include alignment with states of knowledge, and social-psychological aspects of supply chain actors and stakeholders globally. In this section, we map the multiple and simultaneously occurring drivers of deep uncertainty and risk against/along the supply chain uncertainty continuum ('complete certainty' to 'radical uncertainty') to support our argument for other social-psychology aspects to be incorporated in supply chain risk management models.

The literature is filled with advice for managing risks under uncertainty: design fault tolerant, survivable, and resilient organizations, systems, and infrastructure; loosen or decouple the tight couplings and dependencies in existing complex systems and infrastructure; adopt a vigilant, risk-aware mind set, and culture, build highly reliable organizations around the five principles (preoccupation with failure, reluctance to simplify interpretations of data and anomalies, sensitivity to operations, commitment to resilience, and deference to expertise rather than to authority) (e.g., Cox 2012). However, our failure to predict global supply chain disruptions and their impact on local businesses, or the endgame of escalating geo-political tensions between USA and China in the Taiwan Striates demonstrates gaps in current logical and systematic approaches to managing risks under conditions of deep uncertainty. Managing supply chain risks requires broadening the discussion around how risk managers can improve managerial decision making under conditions of supply chain deep uncertainty. It is important to note at this point that the succeeding discussion and recommendations should be viewed as complementary and not contrary to existing quantitative approaches modelled around data, probability and risk assumptions in decision trees, and predication, financial forecasts, and optimisation models, among others.

Managerial decisions and actions during unprecedented conditions vary across the levels of supply chain uncertainty and should be aligned to how different customers and suppliers respond to percieved uncertainty and risks in a multi-tier supply chain. For example, deployment of one strategy by a customer to overcome probabilistic uncertainty related disruptions could create a cascading effect causing procedural uncertainties for 3rd, 4th, and 5th tier suppliers. While several strategies for mitigating uncertainty are highlighted in the literature: postponement, collaboration to minimise the barriers between decoupling points and decision making complexities, automated processes that limit human intervention related uncertainties; and embedding flexibility through multiple sourcing (Simangunsong et al. 2012), these strategies need to be advanced to provide for unprecedented deep supply chain uncertainties from multiple and simultaneous mega stressors. The suggestions below include factors that supply chain risk managers should put into consideration when establishing strategies for mitigating systematic supply chain disruptions.

Through conceptualising the drivers of deep uncertainty and risk along an uncertainty continuum, risk managers can better understand how misconceptions of COVID-19 as an unknown-unknown ('Black Swan') changed how supply chain actors and consumers responded to the COVID-19 pandemic. With knowledge, supply chain risk managers would have pre-empted or at least mitigated the experienced supply chain stockouts and disruptions (see Fig. 4). For example, the fact that the world has experienced previous flu pandemics calls for a de-escalation of COVID-19 from the level of radical uncertainty (unknown-unknowns or Black-swans) to procedural uncertainty: a situation where the supply chain risk manager is constrained by computational and cognitive capabilities to accurately define the risk event due to limited knowledge about the pandemic (Vilko et al. 2014). Risk managers need to assume that another flu pandemic will emerge in the future and, should not be treated as unknown-unknowns (Black Swan). Unlike pandemics, natural disasters, and geo-political tensions, broad supply chain cyberattacks can be placed in the category of probabilistic uncertainty because risk managers have some experience and knowledge about the potential risk events occurring and can attach probabilities and estimate the likely impact, although with a very limited degree of accuracy. For example, broad supply chain cyberattacks cover a large spectrum because of the deep uncertainty posed by the likelihood of a successful indirect cyberattack through 3rd and 4th tier suppliers who are at times not even known to the targeted customer.

Understanding the different states of knowledge. Supply chain risk management models that don't take into consideration the state of knowledge push risk managers to make decisions focused mainly on optimisation while paying limited attention all critical drivers of deep uncertainty. Looking back at the simultaneous occurrence of multiple supply chain stressors, one can argue that hypothetical knowledge states (*Known-Knowns)* may not apply even where a manager has definitive information to pre-empt capacity constraints e.g., in distribution outlet networks (Lee and Kim 2002). In a *Known-Unknows* state of knowledge, a supply chain risk manager needs to clearly identify what information they do not know about drivers of deep supply chain uncertainty when making judgements on which strategy to employ to manage supply chain risks. For instance, a supply chain risk manager may not be able to determine the effect of ICT risk management solutions across different levels in a multi-tier supply chain, or the energy costs to the supply chain system that is inherently digitised and driven by technological function (Horner et al. 2016). However, understanding and providing for this gap of knowledge could make the difference in whether a risk manager makes a good judgement on how to address prevailing supply chain risks and related systematic disruptions. Although the *Unknown-Unknowns state of knowledge* is more of a hypothetical condition (Black Swans Theory), it's important that supply chain risk managers acknowledge outliers and make provisions for responding to sources of risk that are not yet known. The Black Swans Theory explains an event as an outlier, outside the realms of regular expectations where nothing in the past pointed to its possibility. Human nature provides explanations for its occurrence after the fact, making it explainable and predictable (Weber 2021). Supply chain risk managers with this knowledge should be able to switch their managerial decision making processes say from logical-incrementalism to disjointed-incrementalism in order to cope mega sources of risk such as the COVID-19 pandemic that come with unprecedented levels of deep uncertainty and risk.

Managing perceptions and attitudes toward uncertainty and risk. Risk perception is a subjective judgement by managers based on the estimate of the severity or impact of risk events. Subjective judgement of risk is driven by the manager's state of knowledge and varies across the supply chain uncertainty continuum. According to Friday et al. (2021), understanding the factors influencing risk perception (e.g., the scope of the event, controllability, awareness, trust among customers, suppliers and government authorities, and personal impact and experience) is key to establishing how organizations prepare and respond to unprecedented deep uncertainties such as that caused by COVID-19 pandemic or the Russia-Ukraine war. The factors play a key role in shaping the risk attitude of a supply chain risk management and suppliers and the strategic choices they make under conditions of deep uncertainty. Below are hypothetical examples on how managers with varying perceptions of uncertainty and risk respond to supply chain threats:

- *Risk Averse*—A supply chain risk manager who chooses a sure option over a gamble with comparable expected value is risk averse. In this context, a manager may prefer a guaranteed level of profit to a potentially unknown probability of achieving a higher level of profit when weighing up business investment opportunities through contract designing.
- *Risk Seeker*—A supply chain risk manager who chooses a gamble over the sure amount is prone to taking on more risk. Yet, risk seeking managers might accept varying levels of risk across an uncertainty continiuum i.e., a manager may seek specific lelvels of risk given the potential benefits and known or unknown uncertainty.
- *Risk Neutral* is the recommended attitude for supply chain risk managers dealing with unprecedented deep supply chain uncertainty and related risks. It is a position of indifference (risk neutral), in which the supply chain risk manager aims to examine all the facts based on available information and employ strategies with the highest payoff. In this situation, the outcome represents an adequate approach to forecasted investment tradeoffs for profit or loss, where the manager themselves is indifferent to the outcomes. This can be helpfull in contexts of information uncertainty when dealing with economic forces affecting the supply function.

Relying on logic during times of deep supply chain uncertainty is easier said than done. Systematic supply chain disruptions create deep uncertainties that require risk managers to consider so many moving pieces (some outside their knowledge or expertise) in their decisions. Supply chain risk managers can only remain rational under conditions of deep supply chain uncertainty up to a limited extent. Rational choice under uncertainty simply means that there is a 100% chance of achieving the outcome of our choices. In decision making under certainty, risk managers know for certain what the risk is, the available options, and can define the utility derived from each choice. Making the choice takes a little more than a comparison of risk strategies to maximize utility. Thus, if the axioms of decision making under conditions of certainty hold true, then supply chain risk management choices can be ranked from most preferred to least preferred to maximize utility. However, deep uncertainties introduce varying levels of probability that are at times beyond a supply chain risk manager's comprehension. These conditions make supply chain risk managers more susceptible to compromising and trading off rational alternatives for emotional or ethical concerns. However, relying on decision support systems enabled by cutting edge technologies in big data analysis and machine learning (blockchain, artificial intelligence) can ensure that logic is maintained under unprecedented conditions of deep uncertainty. These technologies can ensure that supply chain risk managers select strategies with optimal payoffs using available information about a known predictable risk or an unfolding supply chain disruption.

Awareness of information biases in individual, group, and organisational decision making. It is worth engaging in effortful thought processes to make quality supply chain risk management decisions under conditions of deep uncertainty. The secret to better judgement under conditions of unprecedented uncertainty is in learning the difference between appropriate and inappropriate heuristics (Bazerman and Moore 2017). While the human brain has developed over centuries, risk managers like the rest of us remain ignorant of the internal workings of the mind and how first our thought processes are easily constrained by biases such as ease of recall from immediate memory, and confirmatory hypothesis testing. As such, supply chain risk managers need to pay attention to decision quality compromises when sought information for analysis is meant to confirm pre-determined choices (stipulated in budgets and plans) or fail to adjust mindsets from previously formed value estimates (anchors in re-negotiation of previous/new supply contracts during conditions of unprecedented uncertainty).

Collaboration across government and the private sector. Regardless of the source of disruption, interventions for managing supply chain risks are undermined by their inability to cope with deep uncertainties caused by multiple and simultaneously occurring mega trends whose disruption effects are beyond the comprehension of a risk manager or single firm: rapid supply chain digitalisation, environmental transformations around climate change; genetic engineering; nuclear energy; economic/financial crises; cyber-terrorism; and environmental degradation, (Renn et al. 2011). Contrary to studies arguing that supply chain collaborative arrangements contribute to increasing uncertainty (Chin-Fu et al. 2005), alternative opinions propose collaboration arrangements as a solution because it increases signalling and visibility through capabilities such as information sharing and process integration (Knight 2012; Simangunsong et al. 2012; Vilko et al. 2014). A European Union Agency for Cybersecurity (2021) report on the 'threat landscape for supply chain attacks' re-affirms the importance of collaboration. The report indicates that up to 62% of cyberattacks are a result of exploited relationships with trusted suppliers, and without augmenting external relationships, customers and suppliers will experience an increase in cyberattack related disruptions.

Establishing a supply chain learning curve. The decoupled nature of global supply chains can create hindrances for supply chain actors/agents responding to unprecedented disruptions from geo-political conflicts or the COVID-19 pandemic, from learning from each other. According to Friday et al. (2021), this is because of differences in objectives and variations in learning experiences across private and public sectors, non-governmental organizations (NGOs), developing and developed economies. Compounded by the decoupling inherent in supply chain structures, focus on individualistic objectives undermines the ability of supply chains impacted by mega stressors such as international wars and escalating geo-political tensions to develop and document a learning culture that can be shared to ensure risk managers do not repeat mistakes. For example, in reponding to supply chain risks posed by the COVID-19, risk managers should have relied on examples and knowledge from previous flu pandemics, and experiences from countries impacted epidemics such as Ebola (Liberia, Sierra Leone, Guinea, and Uganda). Knowledge and experiences from these countries were not leveraged in the early days of the global response to COVID-19 related supply chain disruptions due to a lack of accurate documentation in developing countries. As such, there is an urgent need for risk managers and researchers alike to establish models of supply chain learning curves that capture the current unprecedented times to inform how supply chain risk management decisions are arrived at under conditions of deep uncertainty in the future.

7 Conclusions

This chapter brings unprecedented stressors and sources of supply chain deep uncertainty and risk to the attention of researchers and practitioners in supply chain risk management. The focus of the chapter is to assess and explain why risk managers (regardless of industry sector) need to focus on four key points relating to deep uncertainty to make rational judgements in supply chain risk management. The proposed four points include; establishing the broad drivers of deep uncertainty, states of nature (known probabilities), states of knowledge, and the supply chain risks at hand. Drawing from our analysis, we conclude that supply chain deep uncertainty and related risks are compounded by the inherent decoupling, complexity, and spill-over effects from actions, perceptions, and behaviours of every stakeholder in a global supply chain ecosystem. As such, existing logic-based strategies based on scenario planning, forecasting, and optimisation of supply chain processes can only go so far in providing managerial insights (e.g., on inventory estimates, customer and supplier perceptions and risk taking preferences) to enable rational decision making in different conditions of deep uncertainty and risk. Towards this end, we argue that existing risk management models are vulnerable to unprecedented or unknown sources of deep uncertainty and risk such as pandmeics or state-sponsored supply chain cyberattacks, and cannot accurately forecast every business or country's core strategic interests, political processes, national security priorities, and diplomatic capabilities. Yet, integreting of these issues as sources of deep uncertainty in a suply chain risk management process is critical for delivering just about every global supply chain project (Wolf and Kalish 2021). We posit that the complexity of global supply chain structures will only continue to increase, contrary to propositions for reshoring and localisation of manufacturing capacity to build a competitive advantage with only friendly countries post the COVID-19 pandemic or the Russia-Ukraine war. In conclusion, future approaches for mitigating supply chain disruptions will require adjusting traditional supply chain risk management models to capture existing drivers of deep uncertainty such those addressed in this chapter, before proceeding to respond to any prevailing risks or threats.

Acknowledgements This book chapter originates from the PhD thesis of the first author, Dr. Derek Friday, in which he examines the relationship between Collaborative Risk Management and Supply Chain Resilience. A copy of the thesis can be accessed by sending a request to Derek.Friday@waikato.ac.nz or via the NOVA open access repository that showcases the research outputs of postgraduate students and academic staff of the University of Newcastle.

References

Applequist GE et al (2000) Risk and uncertainty in managing chemical manufacturing supply chains. Comput Chem Eng 24(9–10):2211–2222

Bailey T, Barriball E, Dey A, Sankur A (2019) A practical approach to supply chain risk management. McKinsey & Company. https://www.mckinsey.com/business-functions/operations/our-insights/a-practical-approach-to-supply-chain-risk-management. 10 Apr 2022

Bazerman HM, Moore AD (2017) Judgement in managerial decision making, 8th edn. Wiley

Bhattarai A (2022) Delayed raises and renovations: small businesses face new uncertainties. the washington post. https://www.washingtonpost.com/business/2022/08/14/small-business-economic-uncertainty/

Bishopa J, Bell T, Huang C, Ward M (2021) Fire on the farm assessing the impacts of the 2019–2020 bushfires on food and agriculture in Australia. file:///C:/Users/df286/Downloads/WWF%20Report-Fire%20on%20the%20Farm_converted.pdf. Accessed 26 Aug 2022

Caldara D, Conlisk S, Iacoviello M, Penn M (2022) The effect of the war in ukraine on global activity and inflation. In: FEDS notes. Board of Governors of the Federal Reserve System, Washington. https://doi.org/10.17016/2380-7172.3141

Centres for Disease Control and Prevention (2022) U.S Department of Health and Human Services. https://www.cdc.gov/flu/pandemic-resources/basics/past-pandemics.html. Accessed 31 Aug 2022

CDP (2020) Transparency to transformation: a chain reaction—CDP Global supply chainreport 2020. file:///C:/Users/df286/Desktop/CDP_SUPPLY CHAIN_Report_2020%20Supply%20Chain%20Climate%20Change.pdf

Chin-Fu H, Yen-Ping C, Yi-Ming T (2005) A structural approach to measuring uncertainty in supply chains. Int J Electron Commer 9(3):91–114. http://search.ebsupplychainohost.com/login.aspx?direct=true&db=bth&AN=16934965&site=eds-live

Corbeau A (2022) How deep is Europe's dependence on Russian oil? https://news.climate.columbia.edu/2022/03/14/qa-how-deep-is-europes-dependence-on-russian-oil/#:~:text=How%20dependent%20is%20Europe%20on,gas%20demand%20of%20512%20bcm. Accessed 9 Apr 2022

Cox LA Jr (2012) Confronting deep uncertainties in risk analysis. Risk Anal Int J 32(10):1607–1629

ENISA (2021) ENISA threat landsupply chainape for supply chainattacks. European Union Agency for Cybersecurity. https://www.enisa.europa.eu/publications/threat-landsupplychainape-for-supply-chain-attacks. Accessed 16 Sep 2021

Escudero LF et al (1999) Schumann, a modeling framework for supply chain management under uncertainty. Eur J Oper Res 119(1):14–34

Friday D, Savage DA, Melnyk SA, Harrison N, Ryan S, Wechtler H (2021) A collaborative approach to maintaining optimal inventory and mitigating stockout risks during a pandemic: capabilities for enabling health-care supply chainresilience. J Humanit Logist Supply Chain Manag 11(2):248–271. https://doi.org/10.1108/JHLSUPPLYCHAINM-07-2020-0061
Gong J, Mitchell JE, Krishnamurthy A, Wallace WA (2014) An interdependent layered network model for a resilient supply chain. Omega 46:104–116. https://doi.org/10.1016/j.omega.2013.08.002
Guénette J, Wheeler C (2022) World Bank Blogs. The global economic outlook in five charts. https://blogs.worldbank.org/developmenttalk/global-economic-outlook-five-charts. Accessed 26 Aug 2022
Guénette J, Kenworthy P, Wheeler C (2022) Implications of the War in Ukraine for the Global Economy. World Bank Group. https://thedocs.worldbank.org/en/doc/5d903e848db1d1b83e0ec8f744e55570-0350012021/related/Implications-of-the-War-in-Ukraine-for-the-Global-Economy.pdf. Accessed 26 Aug 2022
Gupta A et al (2000) Mid-term supply chain planning under demand uncertainty: customer demand satisfaction and inventory management. Comput Chem Eng 24(12):2613–2621
Helbig C, Gemechu ED, Pillain B, Young SB, Thorenz A, Tuma A, Sonnemann G (2016) Extending the geopolitical supply risk indicator: application of life cycle sustainability assessment to the petrochemical supply chain of polyacrylonitrile-based carbon fibers. J Clean Prod 137:1170–1178
Horner NC, Shehabi A, Azevedo IL (2016) Known unknowns: indirect energy effects of information and communication technology. Environ Res Lett 11(10):103001
Jüttner U, Peck H, Christopher M (2003) Supply chain risk management: outlining an agenda for future research. Int J Log Res Appl 6(4):197–210
King A, Gallagher M (2020) Cyberspace Solium Commission Final Report. United States of America Cyberspace Solarium Commission. https://drive.google.com/file/d/1ryMCIL_dZ30QyjFqFkkf10MxIXJGT4yv/view
Knight FH (2012) Risk, uncertainty and profit: courier corporation
Lee YH, Kim SH (2002) Production–distribution planning in supply chain considering capacity constraints. Comput Ind Eng 43(1–2):169–190
Li X, Li YJ (2022) NEV's supply chain coordination with financial constraint and demand uncertainty. Sustainability 14(3)
MacMahon A, Smith K, Lawrence G (2015) Connecting resilience, food security and climate change: lessons from flooding in Queensland, Australia. J Environ Stud Supply Chainiences 5(3):378–391
Manuj I, Mentzer JT (2008) Global supply chain risk management. J Bus Logist 29(1):133–155
Melnyk SA, Supply chainhoenherr T, Speier-Pero C, Peters C, Chang JF, Friday D (2021) New challenges in supply chainmanagement: cybersecurity across the supply chain. Int J Prod Res
Phillips J, Heldman K, Baca C, Jansen P (2006) PMP: project management professional study guide. McGraw-Hill
Pournader M, Kach A, Talluri S (2020) A review of the existing and emerging topics in the supply chain risk management literature. Decis Sci 51(4):867–919
Prakash S, Prakash S, Soni G, Soni G, Rathore APS, Rathore APS (2017) A critical analysis of supply chainrisk management content: a structured literature review. J Adv Manag Res 14(1):69–90
Raian S et al (2022) Assessing sustainability risks in the supply chain of the textile industry under uncertainty. Resour Conser Recycl 177
Renn O, Klinke A, Asselt M (2011) Coping with complexity, uncertainty and ambiguity in risk governance: a synthesis. AMBIO—J Hum Environ 40(2):231–246
Simangunsong E, Hendry LC, Stevenson M (2012) Supply-chain uncertainty: a review and theoretical foundation for future research. Int J Prod Res 50(16):4493–4523
Statista (2022) Share of the total gross domestic product (GDP) generated by travel and toursim worldwide from 2022 to 2021. https://www.statista.com/statistics/1099933/travel-and-tourism-share-of-gdp/. 26 Aug 2022

Townsend DM, Hunt RA, McMullen JS, Sarasvathy SD (2018) Uncertainty, knowledge problems, and entrepreneurial action. Acad Manag Ann 12(2):659–687

Trkman P, McCormack K (2009) Supply chainrisk in turbulent environments—A conceptual model for managing supply chainnetwork risk. Int J Prod Econ 119(2):247–258

Vacante M, D'Agata V, Motta M, Malaguarnera G, Biondi A, Basile F et al (2012) Centenarians and supercentenarians: a black swan. Emerging social, medical and surgical problems. BMC Surg 12(1):1–8

van der Vorst JGAJ, Beulens AJM (2002) Identifying sources of uncertainty to generate supply chainredesign strategies. Int J Phys Distrib Logist Manag 32(6):409–430

Vilko J, Ritala P, Edelmann J (2014) On uncertainty in supply chain risk management. Int J Logist Manag 25(1):3–19

Wolf M, Kalish I (2021) Supply chainresilience in the face of geopolitical risks—Preparing for the tumult ahead. Deloitte. https://www2.deloitte.com/us/en/insights/economy/us-china-trade-war-supply-chain.html. Accessed 8 Apr 2022

Wong CY et al (2011) The contingency effects of environmental uncertainty on the relationship between supply chain integration and operational performance. J Oper Manage 29(6):604–615

Emergent Technologies for Supply Chain Risk and Disruption Management

Prateek Kumar Tripathi, Arun Kumar Deshmukh, and Tribhuvan Nath

1 Introduction

Disruptions due to the 2011 earthquake and tsunami in Japan had nearly the same impact on supply chains as COVID-19 has had on supply chain performance in various nations (Wharton Business Daily 2020). With the coronavirus-induced shutdown in India, the availability of products on the online channels fell by 10%, along with an equal drop in fruits and vegetables in the wholesale market due to cross-border restrictions and lack of visibility throughout the food chain. Furthermore, the impact of the pandemic was profound on the producers of perishable goods, and where the distance between the supply point and delivery point was farthest (Mahajan and Tomar 2021).

In 2016, the devastating Kaikoura earthquake severely affected the functioning of distributed infrastructures and transportation networks in the southeast region of New Zealand, upending the regular functioning of the product supply chain. Most air and sea networks were unaffected; however, fault rupture and ground shaking posed a substantial risk to road and rail transportation- a major source of food, water, and fuel supply. Albeit, alternative road diversions offered some respite, a sudden increase in the traffic volume led to a delayed post-recovery process (Lui 2016, Nov 13; NZ Herald 2017). Reportedly, in New Zealand, the bottlenecks during and after a disaster in the supply chain of essential items are primarily related to vulnerable infrastructure, transportation networks and lack of real-time information flow. Similarly, severe rains in New South Wales and Queensland, Australia, caused massive-scale destruction

P. K. Tripathi · A. K. Deshmukh (✉)
Institute of Management Studies, Banaras Hindu University, Varanasi 221 005, India
e-mail: akdeshmukh@fmsbhu.ac.in

T. Nath
RGSC, Institute of Management Studies, Banaras Hindu University, Varanasi, India

S. K. Paul et al. (eds.), *Supply Chain Risk and Disruption Management*, Flexible Systems Management, https://doi.org/10.1007/978-981-99-2629-9_4

to communities and businesses in these states in March 2022, leading to a national emergency declaration (Burke 2022; BBC News 2022) and disruption of supply chain processes.

The primary supply chain risks observed in the supply chain network (SCN) of any product are caused by inherent uncertainties (operational risk) and man-made or natural disaster settings (disruption risk). Essentially, organizations of both private and public types, fail to manage unforeseen and known–unknown supply chain risks because of a lack of transparency at the supplier base, difficulty to quantify disruption likelihood, destructive strength, and limited visibility (Bailey et al. 2019). The responsible factors for operational risk are uncertain external events, uncertain demand variability, instability in supply, processing time and asymmetry in information dissemination whereas those for disruption risk are earthquakes, floods, hurricanes, terrorist attacks, and more (Ivanov et al. 2019; Tang 2006). These unpredictable disruptions lead to uncertainty across the supply chain, causing deleterious impacts on its operation by impacting SCN entities, namely, external suppliers, production units, distribution units, demand zones, and logistics assets (Klibi and Martel 2012; Wu et al. 2007). This gets exacerbated by critical nodes, vulnerable infrastructure, inefficient supply chain monitoring, and improper coordination network (Akhtar et al. 2012; Davidson 2006; Goentzel 2020; Noori and Weber 2016). Furthermore, disruptive factors consisting of cross-border restrictions and transportation difficulties have a major consequential impact on the functioning of the food supply chain (Sharma et al. 2021), such as quality degradation, more resource wastage, inefficient distribution, and price gouging by the exploitative middlemen. Disasters such as earthquakes make the demand forecasting process ineffective due to abrupt changes in consumer behavior and requirements (Roberson 2019). For example, during this phase, demand for essential food and healthcare items received an unreasonable spike compared to other ranges of products in these segments.

Ideally, in a time of crisis, the business operations should have the capacity to supply desired goods at the right place, in the right quantity, in the right quality, and at the right time through the right transportation model. This response capacity is a key qualifier for commercial supply chain resilience against disruptive events (Duong and Chong 2020; Goentzel 2020; Haghani and Oh 1996) such as earthquakes, tsunamis, COVID-19, and more. However, in the face of random disruptions, there is no effective contingency plan that can thoroughly recover what is lost in that event; however, having a proactive and agile business model certainly lessens the severity of man-made or natural disasters (Wharton Business Daily 2020; Singh et al. 2021).

Managing supply chain risks, emerging from disruptions caused by the volatile, uncertain, complex, and ambiguous business environment, is feasible through trusted information sharing among supply chain stakeholders (Min 2019; Yadav et al. 2020). Therefore, a robust risk and disruption management operational model would necessitate an accurate assessment of demand disturbance and coordination of supply chain actors with true information exchange (John et al. 2019). Adding to that, a better understanding of disruptive impact leads to quick response time and cost optimization (Wu et al. 2007). Thus, real-time, reliable, and secure information flow on disaster-stricken sites and its impact across supply chain networks become imperative in risk monitoring, network visibility, and the formulation of risk mitigation strategies against disruptive risks. In such scenarios, the informal streaming of real-time data from social networking sites can also be used for a cost-effective and efficient assessment of impact and customer demand. Furthermore, monitoring supplier bases, weather phenomena, and news insights requires more preparation time and strategic mitigation planning in real-time data inflow (Sheffi 2015). However, the primary impediments to information management, needed for demand understanding, network coordination, and impact severity, in the crisis are unreliable and mutable information streams and the unavailability of a shared database (digital ledger), leading to inefficient coordination and collaboration among actors 'Producers-Wholesalers-Retailers-Customers' (Altay and Labonte 2014) and uncontrolled propagation of ripple effects (Christopher and Peck 2004; Dolgui and Ivanov 2021; Hao and Srinath 2020; Kazemi 2019; Meier 2015). Interestingly, emerging technologies: Blockchain Technology (BT), Internet of Things (IoT), Big Data Analytics (BDA), Artificial Intelligence (AI), and Fifth Generation (5G) communication technology can be instrumental in addressing the existing lacuna in supply chain risk and disruption management by its virtue of transparency, distributed network, immutability, scalability, fewer latency, intelligent decision making, and more (Dolgui and Ivanov 2021; Gaur and Gaiha 2020; Ivanov et al. 2019; Min 2019; Sharma et al. 2021; Taboada and Shee 2021).

For instance, how AI integrated with advanced analytics can transform data from varied sources into real-time visibility of values flow across the supply chain; and how 5G-enabled network communication would offer real-time intelligent connectivity of new-age technologies, namely, IoT, AI, and Blockchain. The present study explores multiples gaps in existing literature vis-à-vis practice in terms of application of the aforementioned technologies, the context of its use (such as high-end sophisticated large industry vis-à-vis small and medium-level enterprises), geographic and population gap with regard to the number of studies in emerging markets like India vis-à-vis worldwide, methodological gap pertaining to the majority of studies following mono-method to pinpoint the role these technologies play in different walks of industrial reality, and more studies are desired in this pretext with multi-method triangulation to draw robust conclusion with greater generalizability. As supply chain risk and disruption management have not yet received substantial attention on the intervention of new-age technologies in various contexts, the present study contributes comprehensively to the existing body of knowledge and practices in this domain. The remaining paper presents a review of literature with theoretical background and role of emerging technologies in supply chain risk and disruption management followed by the proposed model, the section on major findings covers a case study and a primary survey-based study on SCRM. The next section revolves around the discussion and implications and sums up with conclusions and direction for future research.

2 Review of the Literature

The present section encompasses specific sections covering theoretical background, and literature review on emerging technology and SCRM with a special focus on the mentioned technologies.

2.1 *Theoretical Background: Boundary Spanning and Innovation Adoption Theory*

The present study is informed by the two prominent theoretical perspectives in organizational literature on how and why enterprises accept the contemporary technologies, which involves boundary spanner theory (Aldrich and Herker 1977; Huang et al. 2016) and innovation adoption theory (Rogers 1995) which was recontextualized earlier by Kimberly and Evanisko (1981) and recently, in the works of Clohessy and Acton (2019).

The seminal research by Aldrich and Herker (1977), the role of boundary spanners was considered to be significant in establishing and executing smooth economic and social exchanges between enterprise and the forces of external environment. Thereby, the spanner facilitates in external representation and information processing (Huang et al. 2016). By performing these roles, the boundary spanners facilitate the information exchange with external entities and cascades the organizational responses to environmental stimuli. According to Markman et al. (2008), the personal ties among the entities in the SCM play a crucial role in building business relationships in emerging market and boundary spanner making business decisions are usually influenced by such personal ties among themselves. Strong ties can help them in boundary spanning by creating a platform for establishing connection and sharing information, and serving as a lubrication for mutual willing cooperation and effective problem-solving, i.e., external representation). Thus, collegial boundary-spanning behavior help the spanner hosting organization to garner favorable long-run relationship among the exchange partners in the supply chain environment (Huang et al. 2016). Contemporary technologies can be considered as facilitating platform for such boundary spanning and thereby, enable them mitigating risk and disruption.

The innovation adoption coupled with boundary-spanning behavioral perspective set the pace of ongoing discussion. Technologies such as blockchain, IoT, AI etc. are considered an IT innovation that is referred to as application of new IT tools by an entity (individual and/or enterprises) (Swanson 2004). Pointing out the peculiarity of IT innovations, Wang (2019) enquired an intriguingly ardent phenomenon on adoption of technology at the organizational level and tried to find reason behind the differing rate of adoption of information technologies. The scholars in the information systems domain considered IT as an organizational innovation and presented several environmental, organizational, and technological (EOT framework) factors contributing to the decision by an organization to adopt and not adopt an innovation (Tornatzky and Fleischer 1990).

Organization's IT adoption can not only take it to the new growth trajectory in the long-run but also lead to significant transformation in its external and internal operation (Wang 2019) and particularly in managing risk and disruptions. It is evident through some of the evolving research on application of artificial intelligence in supply chain risk management (Paul et al. 2020), and big data and blockchain technology in supply chain was studied by Narwane et al. (2021) investigate the how emerging IT solutions can help enterprises mitigate the risk.

Likewise, in UK, a study by Chowdhury et al. (2022) investigated the adoption of blockchain technology for risk management in SCM environment. The implementation of blockchain technology determines reliability, accuracy, timelines, visibility, and precision in supply chain transactions and processes. Consequently, it makes technology lucrative to enhance the transparency, robustness, decision-making, and accountability with respect to risk management. Thus, it is imperative for enterprises prone to high risks and disruptions of complex and uncertain nature to adopt such technologies.

3 Emerging Technologies and SCRM

It is argued that organizations of both private and public types fail to manage unforeseen and known-unknown supply chain risks because of a lack of transparency at the supplier base, difficulty to quantify disruption likelihood and its destructive strength and limited visibility (Bailey et al. 2019). Information processing in uncertain events provides an efficient response to any unwanted deviation from normalcy, as per organizational information processing theory (OIPT) (Dubey et al. 2021). Interestingly, Dubey et al. (2020) confirmed through their study that distributed ledger technology strengthens mutual trust among supply chain actors, and this emerging technology enhances the efficiency and effectiveness of risk and disruption management operations through visibility, accountability, and traceability across the flow of goods. Partner collaboration through blockchain technology networks—the biggest digital disruption among other technological breakthroughs—enhances disruption responsiveness by facilitating real-time information sharing with more transparency, accuracy, traceability, disintermediation, and swift trust-essential for coordination and collaborative efforts—than other contemporary classical technologies (Bosona and Gebresenbet 2013; Chowdhury et al. 2022; Khan et al. 2021; L'Hermitte and Nair 2020; Queiroz et al. 2019; Rejeb et al. 2021; Sharma et al. 2021).

The blockchain ledger can store essential information attributes in an immutable and auditable format, thus improving supply chain performance with high stakeholder engagement, improved creditability and visibility, and reduced value leakages due to counterfeit practices (Gaur and Gaiha 2020; Rogerson and Parry 2020). In addition to that, in a disruptive environment, blockchain acts as a potential driver to identify and control the root cause of ripple effects and provide comprehensive real-time information on capacities and inventories for the formulation of risk mitigation strategies without any need for intermediation. Consequently, the supply chain overcomes operational issues and improves network visibility and responsiveness (Christopher and Peck 2004; Duong and Chong 2020; Min 2019; Sharma et al. 2021) to regulate deviation of any magnitude from normal supply chain operation (Bearzotti et al. 2012; Ivanov et al. 2019). As BT is a highly secure peer-to-peer (P2P) decentralized network of hashed time-stamped blocks, connected through cryptography, it will facilitate decentralization, auditability, and accountability in maintaining standards across the chain in real time and enables proactive risk mitigation and control mechanism (Torky and Hassanein 2020; Yuan et al. 2019) in terms of information transparency, resource, and product visibility (Rogerson and Parry 2020; Treiblmaier 2018). In this regard, Dasan Potty and Yu (2020) quantitatively assessed the impact of blockchain technology on the Walmart transportation service line and found that this technology-driven value chain offered greater visibility for transportation, timely delivery, and low cost of dispute management.

Similarly, the studies were conducted to explore blockchain implications in real-time information traceability, product traceability, smart contract formulation and disintermediation, resource visibility, and addressing agency problem of trust issues (Choi et al. 2019; Lu and Xu 2017; Queiroz et al. 2019; Treiblmaier 2018; Wang et al. 2020). Therefore, it can be decisively proposed that seamless information sharing through a blockchain-based decentralized network entails efficient risk assessment, low response time, accurate market sensing, visibility, and trust (Kumarathunga 2020) establishment.

Integrating blockchain networks with Artificial Intelligence brings about the capacity of real-time data analytics in a Trusted Decentralized Artificial Intelligence (TDAI) system, which could not be possible in the sole application of blockchain technology in supply chain processes (Gulati et al. 2020; Tanwar et al. 2019). Eventually, blockchain implementation would enhance transparency, security, auditability, and trustworthiness of decisions drawn out of AI models in a decentralized fashion. Some of the advantages of Blockchain-enabled AI model entails *data provenance* that ensures the reliability of the data source; thus, decision-making based on AI algorithms would be more effective, *Information intelligence* out of AI-predictive models would be seamlessly disseminated to cross-difference supply chain nodes without any need for performance audit, and *faster and effective real-time analytics* on chained data on account of no data validation requirement (Dinh and Thai 2018; Gulati et al. 2020; Sarpatwar et al. 2019; Tanwar et al. 2019).

Although the Internet of Things (IoT) would bring about the enrichment of data streams through real-time information sharing among different nodes in the supply chain network, entailing quick responsiveness in the chain in the aforementioned AI-enabled BT model in less complex small supply chain networks, in case of complex long networks the issue of low bandwidth, higher latency, low speed necessarily requires the intervention of 5G communication technology. Furthermore, enhanced reliability and security with disruptive wireless 5G technology allow seamless multi-device connectivity in real time at lower cost and energy consumption with higher delivery efficiency (Higgins 2021; Hrouga et al. 2022; Taboada and Shee 2021).

Lastly, large-scale data of both structured and unstructured demand big data analytical processing to garner key decision-driven insights and infuse the right data into AI engines and BT networks (Ivanov et al. 2019).

3.1 Benefits of Emerging Technologies in SCRM

Risk is an integral part of any supply chain but with the application of emerging technologies such as AI/ML/blockchain/IoT/5G, companies have real-time data to identify the potential risk and suggest the best possible mitigation strategies (Barta and Görcsi 2021; Dolgui and Ivanov 2022; Shwetha and Prabodh 2021; Toorajipour et al. 2021; Younis et al. 2021; Tzachor 2020; Baryannisa et al. 2019; Kar et al. 2019; Rejeb et al. 2019; Saberi et al. 2019; Rodrigo et al. 2018; Lamba and Singh 2016). The application of these emerging technologies in SCRM offers multiple benefits to companies. With AI, ML, IoT, blockchain, 5G, and Big Data Analytics, companies can trim unnecessary spending, lower costs, and lower costs and serve customers in a much more effective way. A recent study by McKinsey shows that the companies that have adopted AI solutions reported 15% reduction in logistics costs, improvement in inventory levels by 35% and service levels boost by almost 65%. The supply chain is a connected web or network of multiple functions and functionaries involving procurement, manufacturing, logistics, marketing, and sales. AI tools are applicable throughout the supply chain stages that can minimize risk and increase performance and overall efficiency. The following Table 1 describes the benefits of emerging technologies at different stages of the supply chain, as highlighted by the literature. Table 1: Benefits of AI application in risk management at different stages of supply chain.

4 Proposed Model

In particular, collaboration and coordination among and within primary supply chain actors, namely, producers, wholesalers, and retailers individually through immutable and reliable contracts (smart contracts) facilitate the smooth flow of products through shared logistic support that would not be feasible if all elements of the supply chain, from producers to retailers act independently with their own logistical support. Especially during disruption, this act of collaboration and coordination is indispensable.

Highlighting the importance of individual proposed new-age technologies, cloud computing in a digital supply chain network entails higher responsiveness to supply chain volatility, actionable insights out of high-volume data flow, end-to-end operational visibility, collaboration, and innovative approaches. Additionally, it brings the

Table 1 Benefits of emerging technologies (AI, ML, IoT, blockchain, cloud, 5G, and big data analytics) in risk management at different stages of supply chain (*Source* Compiled by Authors)

Supply chain stages	Benefits of the emerging technologies	Relevant literature
Procurement	Harnessing the disruptive technologies such as AI/ML, IoT, blockchain, and 5G in procurement risk management can provide multiple benefits to the companies. AI-powered procurement software can help procurement professionals to real-time monitoring of various sources of procurement risks such as lead time variability, price fluctuations and poor quality of raw materials. Adoption of these innovative technologies in procurement helps in automating supplier management, selecting the right supplier(s), negotiating the right price, contract award, and much more	Dilmegani (2022), Drew (2022), Allal-Chérif et al. (2021), Komdeur and Ingenbleek (2021), Perera et al. (2021), Rane and Potdar (2021), Wang et al. (2021), Karlsson (2020), Rane and Thakker (2020), Schulze-Horn et al. (2020), Cohen (2018)
Manufacturing	Frontier technologies like AI, IoT, blockchain, and big data analytics take over the manufacturing process by automating repetitive/routine works and processes, cuts downtime, predicts output, improves productivity, and ensures high-quality end products. AI drives the smart maintenance that monitors the productivity of machinery and parts to spot faults, identify potential downtime, improve work efficiency and reduce maintenance costs	Cheng et al. (2022), Santhi and Muthuswamy (2022), Wei and Li (2022), Collins et al. (2021), Helo and Hao (2021), Jha (2021), Kalsoom et al. (2021), Rai et al. (2021), Zuo (2021), Chuprina (2020), Peres et al. (2020), Zhang et al. (2020), Ko et al. (2018)
Logistics	AI, IoT, blockchain, data analytics, cloud, and 5G are the foundations of Industry 4.0. These promising technologies are most commonly used by logistics companies to minimize risk and reduce logistics cost. AI offers logistics managers real-time data and information, which can be used by managers in moving materials and other resources from one place to storage at the desired destination. Automation is perhaps the most important benefit of using AI in logistics	Boute and Udenio (2021), Santhi and Muthuswamy (2022), Yang and Huang (2021), Min (2010), Musigmann et al. (2020), Pasonen (2020), Aguezzoul and Pires (2019), Hellingrath and Lechtenberg (2019), Pandian (2019), Klumpp (2018)

(continued)

Table 1 (continued)

Supply chain stages	Benefits of the emerging technologies	Relevant literature
Transportation	AI with other disruptive technologies like IoT, blockchain, and 5G are providing a smarter and brighter horizon for the transportation industry. These innovative technologies help in managing transportation risk through vehicle route optimization, tracking road traffic, real-time tracking of lead time status, shipping volume prediction, reduce fuel consumption, and improve planning deliveries	Maqbali et al. (2021), Humayun et al. (2020), Muzylyov and Shramenko (2020), Abduljabbar et al. (2019), Wang and Qu (2019), Zhang (2019), Klumpp and Ruiner (2018), Rymarczyk and Kosowski (2018)
Warehouses management	AI and IoT drive to track warehouse operations in real time for better insight and effective risk management that lead to improve warehouse performance by streamlining processes. AI empowers warehouse automation including using robots to improve the productivity in pick-and-pack processes. AI-powered warehouses can help managers in manpower planning for different warehouses including identifying the best transport options required for different materials and best available routes	Fatima et al. (2022), Shao et al. (2022), Alangari and Khan (2021), Yang et al. (2021), Du (2020), Koricanac (2020), Pandian (2019), Rymarczyk and Kłosowski (2018), Thamer et al. (2018)
Inventory management	Leveraging next-generation technologies such as AI, IoT, blockchain, and 5G in inventory management provides powerful insights for companies such as analyzing factors like weather conditions, market trends, accurate prediction of demand, determining re-order points (ROP), economic order quantities (EOQs), optimize stocks, raise purchase orders, and much more	Osman et al. (2022), Preil and Krapp (2022), Ahmadi et al. (2021), Di (2022), Jondhale and Khairnar (2020), Lingam (2018), Wang et al. (2018), Šustrová (2016)
Sales forecasting	Application of transformation technologies like AI and Big Data Analytics accelerate sales due to much accurate sales' forecasts and better managerial insights that lead to improving the overall sales process	Rohaan et al. (2022), Modgil et al. (2021), Sohrabpour et al. (2021), Wang (2021); Mahroof (2019)
Managing distribution centers	AI/ML-enabled distribution centers can create faster and safer operations that improve customer serviceability	

advantage of 30% lower energy consumption and carbon emission than the internal legacy technology (Accenture 2014). Hence, the entire operation of the suggested modern digital technologies is proposed to be cloud-based (Fig. 1).

To illustrate the function of the proposed AI-enabled Blockchain Network Model, let us assume that C1, C2, C3, C4, … are communication platforms or market sensing units such as social media and more with or without IoT devices to deliver information

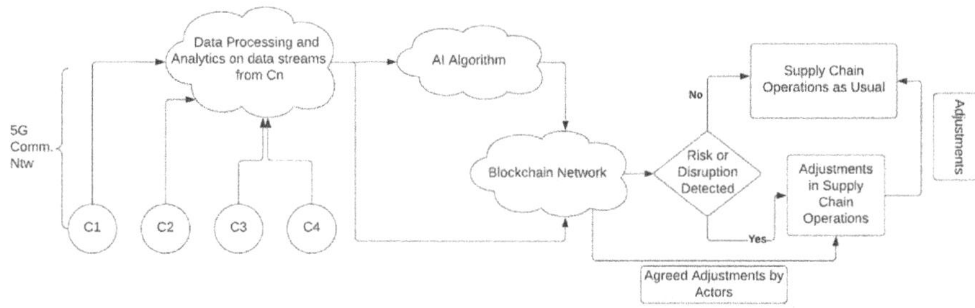

Fig. 1 Functional flow diagram of the proposed framework (*Source* Authors)

of risk or disruptive information. These points, under 5G network, generate large-scale data streams, in a structured or unstructured format, from varied sources such as demand centers, disaster-stricken sites, transportation/logistical modes, wholesalers, retails, producers, and others. With the help of Big Data Analytical Tools (Apache Hadoop; Apache Spark, etc.), this large data volume can be converted into actionable insights to be broadcasted on the blockchain network or curated for AI operation to derive further meaningful predictive or prescriptive insights.

Blockchain network—preferably hybrid to have inter-organizational authentic connect—can be built from Blockchain as a Service (BaaS) providers such as Amazon Managed Blockchain to avoid complex development process of blockchain application network. Notably, the Blockchain ecosystem is a complex network structure that needs adequate planning for managing disruption because not all types of data should be dumped on the blockchain network. Otherwise, its performance would go down; consequently, its potential benefits would turn out to be a loss-making business in risk control and mitigation strategy. Thus, only relevant data from analytics as well as a predictive recommendation from AI engines should be deployed among participants, connected with smart contracts, in the distributed peer-to-peer network. In this way, supply chain key actors would have transparent dashboard view of resources, information, and demand movement; and they would move the supply chain operation, even in a disruptive circumstance, in a coordinated fashion efficiently. In particular, collaboration and coordination among and within primary supply chain actors, namely, producers, wholesalers, and retailers, individually through immutable and reliable contracts (smart contracts) facilitate the smooth flow of products through shared logistic support that would not be feasible if all elements of the supply chain, from producers to retailers act independently with their own logistical support. Especially during disruption, this act of collaboration and coordination is indispensable.

5 Major Findings

5.1 Study I: Case Study: Emerge Stronger at a Time of Uncertainty: Blockchain for Supply Chain (Forrester 2020)

The coronavirus pandemic has reinforced the need for supply chain risk and disruption management that had received relatively little research interest before its onset. Witnessing supply chain vulnerabilities has prompted many businesses to reconsider their operational model and redesign the process to absorb the shocks of disruptions. With this in mind, many are deliberating on tapping the potential benefits of emerging digital technologies, namely, Blockchain, Internet of Things (IoT), Cloud Computing, and more to enhance data quality, integrity, and visibility that are inarguably the necessary assets to address uncertainty confidently in real time.

In a survey conducted by Forrester Consulting and commissioned by IBM, supply chain decision-makers (base: 150) claimed that COVID-19 is among the other seven disruptors (Suppliers going out of business; Price/Currency fluctuation; Transportation failures; Unplanned technical or communications outages; Product problems; Cyberattacks, and new regulations) being the biggest disruptor of all time in the past 12 months—77%. Among other ongoing disruption factors, transportation failures and unplanned communication outages were responsible for 23% each. In addition, reportedly supply chain decision-makers (the same respondents) also grapple with planning challenges due to little to no network visibility due to chain disruption.

Timely access to trustworthy supply chain network data is the pre-requisite for risk and disruption management and that needs to shift to Digital Supply Chain Operations from the traditional style. The survey respondents (59%) endorse supply chain digitization in the next few years on a priority basis. Although real-time data stream can be offered by other new-age technologies such as IoT, 5G Communication, Cloud Computing, and others, due to inaccurate data flow the whole data insights drawn can be of no avail.

Interestingly, distributed ledger technology (Blockchain) has the potential to add the element of trusted data to the aforementioned data stream; however, it is still at a nascent stage in many supply chain businesses. Blockchain implementation entails multiparty data around true data and also brings transparency within and outside business operations with its distributed presence.

Conclusively, 79% of respondents reported transformational positive changes, resulting in reduced risk, improved flexibility and sustainability as well as augmented responsiveness. This demonstrably substantiates the strategic use of emerging technologies, especially Blockchain Technology.

Table 2 Perceived benefits of AI in SCRM (*Source* Compiled by Authors from survey data)

Supply chain stages	Mean*	Mode	Standard deviation
AI for procurement risk management	4.0	4	0.880
AI for manufacturing risk management	4.1	4	0.856
AI for logistics risk management	4.3	5	0.950
AI for transportation risk management	3.9	4	1.058
AI for warehousing risk management	4.2	5	0.987
AI for inventory risk management	4.2	5	1.099
AI for sales forecasting	4.2	4	0.954
AI for managing distribution centers	4.0	5	0.999

* Strongly disagree—1, ..., strongly agree—5

5.2 *Study II: Perceived Benefits of AI in Supply Chain Risk Management*

To get managerial insight on the perceived benefits of AI in SCRM by companies in India, an online survey was conducted in 2022 of conveniently selected 32 manufacturing and service sector organizations located in India. The survey also examined status of presently use of AI, 5-year adoption rate and 10-year adoption rate for SCRM. A survey questionnaire was developed considering the study objectives. A 5-point Likert-type scale was used to express how much they agree or disagree with a particular statement relating to the perceived benefits of AI in SCRM. The collected data was examined using the SPSS 20.0. A descriptive analysis was carried out and simple statistics such as frequency, mean and standard deviation were attempted.

This analysis is presented in Table 2. The results indicate that managers and executives have shown a positive attitude towards the benefits of AI in managing risk in all the stages of supply chain. This denotes that future supply chains will be powered by AI to provide more efficient and risk-free operations. The future companies would be investing more to digitizing their factories and automate operations such as procurement, manufacturing, and logistics to optimize their resource use and maximize benefits. Using AI will enable them to obtain accurate sales forecasts and the highest customer serviceability.

5.3 *Present Adoption of AI in Supply Chain Risk Management*

Though, adoption of AI in supply chain risk management offers many benefits in terms of cost saving, resource savings, time-saving, and increased profit but at the same its adoption by companies is still low. Our research findings show that the highest 71.9% of participating companies use AI for sales forecasting followed by

Table 3 Presently, use of AI in SCRM (*Source* Compiled by Authors from survey data)

Supply chain stages	Yes	No
AI for procurement	53.1% (17)	46.9% (15)
AI for manufacturing	43.8% (14)	56.2% (18)
AI for logistics	62.5% (20)	37.5% (12)
AI for transportation	43.8% (14)	56.2% (18)
AI for warehousing	50.0% (16)	50.0% (16)
AI for inventory management	53.1% (17)	46.9% (15)
AI for sales forecasting	71.9% (23)	28.1% (9)
AI for managing distribution centers	50.0% (16)	50.0% (16)

In bracket corresponding frequency

62.5% for logistics operations, 53.1% for procurement and inventory management, and least 43.1% for manufacturing and transportation operations. About 50.0% of the companies reported using AI to manage distribution centers. In general, a moderate or low adoption of the AI in different stages of supply chain has been observed. This low adoption may be due lack of AI expertise or lack of AI technology. Others limiting factors may be due to budget constraints and difficult to estimate the ROI (Table 3).

5.4 *Future Adoption of AI in Supply Chain Risk Management*

The adoption of AI in SCRM is continuing its steady rise. AI-based solutions are benefiting the companies in managing supply chain risks and improving performance. AI is expected that AI will be the most promising technology for supply chain stakeholders in the coming years. Our research findings are in the same line that shows the company's future move in terms of projected 5-year and 10-year adoption rate of AI in SCRM. According to the results presented in Table 4, the projected 5-year adoption rate of AI was highest (93.8%) for sales forecasting followed by 90.6% for manufacturing, 87.5% for logistics and inventory management, 84.4% for procurement and managing distribution centers, 81.2% for warehousing, and least 78.1% for transportation.

A considerable rise in the projected 10-year adoption rate of AI in SCRM can be seen in Table 4. Leveraging AI to its maximum potential was observed in procurement, manufacturing, logistics, inventory management, sales forecasting, and managing distribution centers with more than 90.0% adoption rate of AI in projected 10-year. Relatively, a low adoption rate was observed for transportation (84.4%) and warehousing (87.5%).

The above findings indicate that AI-powered solutions in upcoming years will be available, affordable, and accessible to help companies to SCRM in order to attain

Table 4 Projected 5-year and 10-year adoption rate of AI in SCRM (*Source* Compiled by Authors from survey data)

Supply chain stages	Projected 5-year adoption rate		Projected 10-year adoption rate	
	Yes	No	Yes	No
AI for procurement	84.4% (27)	15.6% (5)	90.6% (29)	9.4% (3)
AI for manufacturing	90.6% (29)	9.4% (3)	90.6% (29)	9.4% (3)
AI for logistics	87.5% (28)	12.5% (4)	93.8% (30)	6.2% (2)
AI for transportation	78.1% (25)	21.9% (7)	84.4% (27)	15.6% (5)
AI for warehousing	81.2% (26)	18.8% (6)	87.5% (28)	12.5% (4)
AI for inventory management	87.5% (28)	12.5% (4)	93.8% (30)	6.2% (2)
AI for sales forecasting	93.8% (30)	6.2% (2)	93.8% (30)	6.2% (2)
AI for managing distribution centers	84.4% (27)	15.6% (5)	96.9% (31)	3.1% (1)

In brackets, the corresponding frequency

the goal of efficiency and effectiveness in all operations. However, considering future demand of AI professionals in industry, the universities/institutions need to launch AI-based programs to prepare industry-ready professionals.

6 Discussion and Implications

The study intended to explore the role of emergent technologies in SCRM and offers several pathbreaking theoretical as well as practical implications. The first theoretical implications emanated from the ongoing discussion on supply chain and disruption debate that erupted after COVID-19, which posed an unknown and uncertain type of risk to the supply chain partners in various industry verticals. The study explicated that managing such risks and disruption and enhancing the supply chain resilience necessitates the use of technologies such as BCT, AI, IoT, 5G communications, and BDA on case-to-case basis. However, in practice, it is much challenging for small enterprises particularly in emerging economy context, to plan high capital-intensive technologies. Moreover, the innovation adoption perspective also guides the adoption of such emerging IT tools with a variety of applications. Second, the role of the boundary spanner could be crucial as an information disseminator in the supply chain environment, which was presented through the study. It appears to be more of a behavioral phenomenon, and thus necessitates collaborative relationships among the chain partners to ensure seamless flow of information and real-time problem-solving (external representation part of boundary spanning theory). Additionally, the blended method approach in the present research not only cross-validated the results but also ensured the methodological triangulation for drawing better conclusion.

Improved resiliency and reactive power against known risks require documenting prospective risks at every possible node in the supply chain based on their severity, their likelihood, and risk preparedness along with their regular monitoring and use of digital technology which aligns well with the boundary-spanning perspective where technology acts as a platform (Aldrich and Herker 1977; Huang et al. 2016) for information sharing and risk and disruption related problem-solving. On the other hand, unknown risks do not have any structured mitigation framework except for establishing multiple defensive layers and promoting an organization-wide risk-awareness culture to dampen the severity of damage associated with them (Bailey et al. 2019; Thomas 2020; Treiblmaier 2018). Arguably, the scope of emerging technologies in supply chain ripple-effect appeasement and resilience is a promising research area to study how these digital technologies in an integrated framework can be leveraged in risk and disruption management (Dolgui and Ivanov 2021) to offer real-time information traceability, product traceability, resource visibility (Lu and Xu 2017; Rogerson and Parry 2020; Treiblmaier 2018; Wang et al. 2020).

Resilience and agility of the supply chain process against both known and unknown disasters is the need of the hour because the vulnerability impacts the availability of products not only locally but also internationally. Citing the case of the Russia-Ukraine conflict to support the aforementioned view, both accounts for 50% and 30% of the world's sunflower oil and wheat supply respectively and some African and least developed countries are considerably dependent on their export. However, unfortunately, they are on the cusp of a hunger meltdown due to no trade out of these war-stricken countries. Likewise, the case of COVID-19, bush fire and flood in Australia, a deadly quack in New Zealand, looming climate crisis, and more are giving impetus to research studies in the field of supply chain risk and disruption management to have robust recovery responsiveness (Goentzel 2020; UN News 2022; Xu et al. 2020).

The proposed framework of integrated strength of emerging and innovative technologies: BT, IoT, AI, 5G communication, and BDA entails risk-preparedness and improved resilience against any disruptive events through a trustworthy peer-to-peer distributed network with high visibility, low latency, and fastest speed of data transmission with the governance of intelligence data-driven insights. In order to harvest these benefits, many enterprises from both manufacturing and service sectors including all types (small, medium, and large) are keen and planning to adopt these disruptive technologies in the coming 5–10 years, as evident from the survey data. Although this field of study is still at an embryonic stage, the consolidated strength of each underlying digital technology makes it the right investment to make.

7 Conclusion and Future Research Direction

As far as academic research literature is concerned, the study would contribute to the existing knowledge base on the role of new-age digital technologies in the supply chain risk and disruption management domain by demonstrating its implication for supply chain coordination, information management and strengthening of demand sensing elements. For industry practices in this area, it will provide a detailed overview of the technology-based risk mitigation framework along with the determinants of high responsiveness to disruption to address disruptive issues of blocked inventory movement, ripple effects propagation, and low transparency. Additionally, the application of the research results should address the pressing need for a resilient supply chain of essential items such as food, fuel, water, medicines, and healthcare products against disruptions and disasters caused by man-made or natural disasters. However, the respective solution capabilities of discussed emerging technologies should be understood appropriately to yield the potential transformational benefits sustainably; otherwise, the whole operation would only get complicated further.

The adoption of information technology tools at the enterprise level has been an emerging phenomenon and cannot be thoroughly investigated using the existing theoretical perspectives such as innovation adoption theory which is ideally inclined toward end-user adoption. Therefore, future studies may be delved deeper to explore the contextual differences and required corresponding constructs that are specific to business-to-business or B2B contexts. More so, a mixed theoretic perspective may further be extended to empirical and positivist approach to establish the findings and enhance the generalizability. Similarly, despite a blended method perspective in the present study, more such studies may be designed where the context is deeply rooted in specific sectoral and industrial setting such as agriculture, heavy industry, consumer goods, services and the like. Such sectoral specifications will enable researchers to identify the cross-context differences and corresponding strategies to deal with them. The emergent information technology solutions give an impetus to build responsive, resilient, and robust supply chain, which paves the way for recuperating to normalcy immediately after disruption.

Acknowledgements The authors thankfully acknowledge the generous support extended for the research from Institute of Eminence (IoE), Banaras Hindu University to promote research work.

Author Declaration The authors share no conflict of interest in the study.

References

Abduljabbar et al (2019) Applications of artificial intelligence in transport: an overview. Sustainability 11

Accenture (2014) Supply chain management in the cloud. https://www.accenture.com/_acnmedia/accenture/conversion-assets/dotcom/documents/global/pdf/dualpub_1/accenture-supply-chain-management-in-the-cloud.pdf. Accessed 3 Mar 2022

Aguezzoul A, Pires S (2019) Use of artificial intelligence in supply chain management practices and 3PL selection. Syst Cybern Inform 17(4):1–12

Ahmadi et al (2021) Using blockchain technology to extend the vendor managed inventory for sustainability. J Constr Mater 3:1–5

Akhtar P, Marr NE, Garnevska EV (2012) Coordination in humanitarian relief chains: chain coordinators. J Humanit Logist Supply Chain Manag

Alangari S, Khan NA (2021) Artificially intelligent warehouse management system. Asian J Basic Sci Res 3(3):16–24

Aldrich H, Herker D (1977) Boundary spanning roles and organization structure. Acad Manag Rev 2(2):217–230
Allal-Chérif O, Simón-Moya V, Ballester ACC (2021) Intelligent purchasing: how artificial intelligence can redefine the purchasing function. J Bus Res 124:69–76
Altay N, Labonte M (2014) Challenges in humanitarian information management and exchange: evidence from Haiti. Disasters 38(1):50–72
Bailey T, Barridall E, Dey A, Sankur A (2019) A practical approach to supply-chain risk management. McKinsey & Company. https://www.mckinsey.com/business-functions/operations/our-insights/a-practical-approach-to-supply-chain-risk-management. Accessed 29 Sep 2021
Barta G, Görcsi G (2021) Risk management considerations for artificial intelligence business applications. Int J Econ Bus Res 21(1):87–106
Baryannisa G, Validi S, Dani S, Antoniou G (2019) Supply chain risk management and artificial intelligence: state of the art and future research directions. Int J Prod Res 57(7):2179–2202
BBC News (2022) Australia floods: PM Morrison to declare a national emergency. https://www.bbc.com/news/world-australia-60672065. Accessed 15 Mar 2022
Bearzotti LA, Salomone E, Chiotti OJ (2012) An autonomous multi-agent approach to supply chain event management. Int J Prod Econ 135(1):468–478
Bosona T, Gebresenbet G (2013) Food traceability as an integral part of logistics management in food and agricultural supply chain. Food Control 33(1):32–48
Boute RN, Udenio M (2021) AI in logistics and supply chain management. https://ssrn.com/abstract=3862541. https://doi.org/10.2139/ssrn.3862541
Burke H (2022) Why it could rain for months on Australia's east coast. ABC News. https://www.abc.net.au/news/2022-03-05/when-will-the-rain-stop/100883168. Accessed 21 Mar 2022
Cheng et al (2022) 5G in manufacturing: a literature review and future research, Int J Adv Manuf Technol
Choi TM, Wen X, Sun X, Chung SH (2019) The mean-variance approach for global supply chain risk analysis with air logistics in the blockchain technology era. Transp Res Part E: Logist Transp Rev 127:178–191
Chowdhury S, Rodriguez-Espindola O, Dey P, Budhwar P (2022a) Blockchain technology adoption for managing risks in operations and supply chain management: evidence from the UK. Ann Oper Res 1–36
Christopher M, Peck H (2004) Building the resilient supply chain. Int J Logist Manag 15
Chuprina R (2020) AI and machine learning in manufacturing: the complete guide. https://spd.group/machine-learning/ai-and-ml-in-manufacturing-industry/
Clohessy T, Acton T (2019) Investigating the influence of organizational factors on blockchain adoption: an innovation theory perspective. Ind Manag Data Syst 119(7):1457–1491
Cohen T (2018) Procurement powered by AI: opportunities, risks, and challenges. https://aibusiness.com/document.asp?doc_id=760726
Collins et al. (2021). Artificial intelligence in information systems research: a systematic literature review and research agenda. Int J Inf Manag 60:1–17
Dasan Potty VK, Yu Z (2020) Increasing supply chain visibility by incentivizing stakeholders to use blockchain
Davidson AL (2006) Key performance indicators in humanitarian logistics. Doctoral dissertation. Massachusetts Institute of Technology
Di H (2022) WITHDRAWN: logistics management inventory model based on 5G network and internet of things system. Microprocess Microsyst 103429
Dilmegani C (2022) AI procurement: why it matters. Applications & Use cases. https://research.aimultiple.com/ai-procurement/
Dinh TN, Thai MTJC (2018) AI and blockchain: a disruptive integration. Computer 51(9):48–53. https://doi.org/10.1109/MC.2018.3620971
Dolgui A, Ivanov D (2021) Ripple effect and supply chain disruption management: new trends and research directions

Dolgui A, Ivanov D (2022) 5G in digital supply chain and operations management: fostering flexibility, end-to-end connectivity and real-time visibility through internet-of-everything. Int J Prod Res 60(2):442–451
Drew M (2022) Impact of AI on the evolution of the procurement & supply chain organizations. https://www.aitimejournal.com/@melissa.drew/impact-of-ai-on-the-evolution-of-the-procurement-supply-chain-organizations
Du X (2020) Research on the artificial intelligence applied in logistics warehousing. In: 2nd international conference on artificial intelligence and advanced manufacture
Dubey R, Gunasekaran A, Bryde DJ, Dwivedi YK, Papadopoulos T (2020) Blockchain technology for enhancing swift-trust, collaboration and resilience within a humanitarian supply chain setting. Int J Prod Res 58(11):3381–3398
Dubey R, Gunasekaran A, Childe SJ, Fosso Wamba S, Roubaud D, Foropon C (2021) Empirical investigation of data analytics capability and organizational flexibility as complements to supply chain resilience. Int J Prod Res 59(1):110–128
Duong LNK, Chong J (2020) Supply chain collaboration in the presence of disruptions: a literature review. Int J Prod Res 58(11):3488–3507
Fatima et al (2022b) Production plant and warehouse automation with IoT and industry 5.0. Appl Sci 12(4):2053
Forrester (2020) Emerge stronger at a time of uncertainty: blockchain for supply chain. Forrester Research, Inc. https://www.ibm.com/downloads/cas/JX9KDGPJ. Accessed 04 Apr 2022
Gaur V, Gaiha A (2020) Building a transparent supply chain blockchain can enhance trust, efficiency, and speed. Harv Bus Rev 98(3):94–103
Goentzel J (2020) Resilient supply chains are crucial to maintaining the consistent delivery of goods and services to the American people. https://ctl.mit.edu/sites/ctl.mit.edu/files/2020_03_17_Webinar_Strengthening_Supply_Chain_Resilience.pdf. Accessed 20 May 2021
Gulati P, Sharma A, Bhasin K, Azad C (2020) Approaches of blockchain with AI: challenges & future direction. In: Proceedings of the international conference on innovative computing & communications (ICICC) 2020, New Delhi, India. https://doi.org/10.2139/ssrn.3600735
Haghani A, Oh SC (1996) Formulation and solution of a multi-commodity, multi-modal network flow model for disaster relief operations. Transp Res Part A: Policy Pract 30(3):231–250
Hao AQ, Srinath S (2020) Humanitarian assistance for markets in conflict: a system dynamics approach
Hellingrath B, Lechtenberg S (2019) Applications of artificial intelligence in supply chain management and logistics: focusing onto recognition for supply chain execution: bridging the gap between information systems research and practice. In: The art of structuring, pp 283–296
Helo P, Hao Y (2021) Artificial intelligence in operations management and supply chain management: an exploratory case study. Prod Plan Control 1–18
Higgins M (2021) Emerging technologies in supply management to consider investing. In: Forbes. https://www.forbes.com/sites/forbestechcouncil/2021/09/08/emerging-technologies-in-supply-management-to-consider-investing-in/?sh=7ce8a91b2043. Accessed 27 Sep 2021
Hrouga M, Sbihi A, Chavallard M (2022) The potentials of combining blockchain technology and internet of things for digital reverse supply chain: a case study. J Clean Prod 130609
Huang Y, Luo Y, Liu Y, Yang Q (2016) An investigation of interpersonal ties in interorganizational exchanges in emerging markets: a boundary-spanning perspective. J Manag 42(6):1557–1587
Humayun M, Jhanjhi N, Hamid B, Ahmed G (2020) Emerging smart logistics and transportation using IoT and blockchain. IEEE Internet Things Mag 3(2):58–62
Ivanov D, Dolgui A, Sokolov B (2019) The impact of digital technology and industry 4.0 on the ripple effect and supply chain risk analytics. Int J Prod Res 57(3):829–846
Jha AK (2021) Artificial intelligence (AI) in manufacturing. Int J Innov Res Eng Multidiscip Phys Sci 9(3):155–160
John L, Gurumurthy A, Soni G, Jain V (2019) Modelling the inter-relationship between factors affecting coordination in a humanitarian supply chain: a case of Chennai flood relief. Ann Oper Res 283(1):1227–1258

Jondhale NS, Khairnar DT (2020) An analytical study of use of an artificial intelligence in inventory management with reference to medium scale manufacturing industries in Nashik industrial estate. Vidyabharati Int Interdiscip Res J 11(1):212–218
Kalsoom et al. (2021a) Impact of IoT on manufacturing industry 4.0: a new triangular systematic review. Sustainability 13(22):12506
Kar U, Dash R, McMurtrey M, Rebman C (2019) Application of artificial intelligence in automation of supply chain management. J Strateg Innov Sustain 14(3)
Karlsson F (2020) The opportunities of applying artificial intelligence in strategic sourcing. Master Thesis TRITA-ITM-EX 2020:245
Kazemi N (2019) Optimal inventory model for managing demand-supply mismatches for perishables with stochastic supply. Doctoral dissertation. Massachusetts Institute of Technology
Khan M, Imtiaz S, Parvaiz GS, Hussain A, Bae J (2021) Integration of internet-of-things with blockchain technology to enhance humanitarian logistics performance. IEEE Access 9:25422–25436
Kimberly JR, Evanisko MJ (1981) Organizational innovation: the influence of individual, organizational, and contextual factors on hospital adoption of technological and administrative innovations. Acad Manag J 24(4):689–713
Klibi W, Martel A (2012) Scenario-based supply chain network risk modeling. Eur J Oper Res 223(3):644–658
Klumpp M (2018) Automation and artificial intelligence in business logistics systems: human reactions and collaboration requirements. Int J Log Res Appl 21(3):224–242
Klumpp M, Ruiner C (2018) Regulation for artificial intelligence and robotics in transportation, logistics, and supply chain management: background and developments. Netw Ind Quart 20(2):3–7
Ko T, Lee J, Ryu D (2018) Blockchain technology and manufacturing industry: real-time transparency and cost savings. Sustainability 10(11):4274
Komdeur EFM, Ingenbleek TMP (2021) The potential of blockchain technology in the procurement of sustainable timber products. Int Wood Prod J 12(4):249–257
Koricanac I (2020) Impact of AI on the warehousing industry in the U.S. smart warehouses. In: XVII international symposium 'Business and Artificial Intelligence', SymOrg 2020. https://ssrn.com/abstract=3802116
Kumarathunga M (2020) Improving farmers' participation in agri supply chains with blockchain and smart contracts. In: 2020 seventh international conference on software defined systems (SDS). IEEE, pp 139–144
Lamba K, Singh SP (2016) Big data analytics in supply chain management: some conceptual frameworks. Int J Autom Logist 2(4):279–293
L'Hermitte C, Nair NKC (2020) A blockchain-enabled framework for sharing logistics resources during emergency operations. Disasters
Lingam YK (2018) The role of artificial intelligence (AI) in making accurate stock decisions in E-commerce industry. Int J Adv Res Ideas Innov Technol 4(3):2281–2286
Lu Q, Xu X (2017) Adaptable blockchain-based systems: a case study for product traceability. IEEE Softw 34(6):21–27
Lui K (2016) A second powerful earthquake has hit New Zealand's South Island. Time. https://time.com/4569301/second-earthquake-new-zealand-kaikoura-wellington/. Accessed 23 May 2021
Mahajan K, Tomar S (2021) COVID-19 and supply chain disruption: evidence from food markets in India. Am J Agr Econ 103(1):35–52
Mahroof K (2019) A human-centric perspective exploring the readiness towards smart warehousing: the case of a large retail distribution warehouse. Int J Inf Manag 45:176–190
Markman GD, Siegel DS, Wright M (2008) Research and technology commercialization. J Manage Stud 45(8):1401–1423
Maqbali KHA, Slimi Z, Balsa A (2021) The Pros and Cons of artificial intelligence use in the logistics sector in Oman. Euro J Bus Manag Res 6(4):197–208

Meier P (2015) Digital humanitarians: how big data is changing the face of humanitarian response. CRC Press
Min H (2010) Artificial intelligence in supply chain management: theory and applications. Int J Log Res Appl 13(1):13–39
Min H (2019) Blockchain technology for enhancing supply chain resilience. Bus Horiz 62(1):35–45
Modgil S, Singh RK, Hannibal C (2021) Artificial intelligence for supply chain resilience: learning from Covid-19. Int J Logist Manag
Musigmann B, von der Gracht H, Hartmann E (2020) Blockchain technology in logistics and supply chain management—A bibliometric literature review from 2016 to January 2020. IEEE Trans Eng Manag 67(4):988–1007
Muzylyov D, Shramenko N (2020) Blockchain technology in transportation as a part of the efficiency in industry 4.0 strategy. In: Advanced manufacturing processes. InterPartner 2019. Lecture Notes in Mechanical Engineering. Springer, Cham
Narwane VS, Raut RD, Mangla SK, Dora M, Narkhede BE (2021) Risks to big data analytics and blockchain technology adoption in supply chains. Ann Oper Res 1–36
Noori NS, Weber C (2016) Dynamics of coordination-clusters in long-term rehabilitation. J Humanit Logist Supply Chain Manag
NZ Herald (2017) True damage of 7.8 Kaikoura quake revealed, and could change earthquake research worldwide. https://www.nzherald.co.nz/nz/true-damage-of-78-kaikoura-quake-revealed-and-could-change-earthquake-research-worldwide/PSIDTEH7JQFPNY3TZOVJ5 7S52E/. Accessed 21 May 2021
Osman BM, Alinkeel S, Bhavshar D (2022) A study on role of artificial intelligence to improve inventory management system. Int Res J Moderniz Eng Technol Sci 4(3):226–233
Pandian AP (2019) Artificial intelligence application in smart warehousing environment for automated logistics. J Artif Intell Capsul Netw 1(2):63–72
Pasonen P (2020) The use of artificial intelligence in the supply chain management in Finnish large enterprises. Master's thesis. University of Vaasa
Paul SK, Riaz S, Das S (2020) Organizational adoption of artificial intelligence in supply chain risk management. In: International working conference on transfer and diffusion of IT. Springer, Cham, pp 10–15
Perera S, Nanayakkara S, Weerasuriya T (2021) Blockchain: the next stage of digital procurement in construction. Academia Letters, proof
Peres et al (2020) Industrial artificial intelligence in industry 4.0—Systematic review. Challeng Outlook. IEEE 8:220121–220139
Preil D, Krapp M (2022) Artificial intelligence-based inventory management: a Monte Carlo tree search approach. Ann Oper Res 308 (17)
PwC Report: an introduction to implementing AI in manufacturing. Global Manufacturing and Industrialisation Summit (GMIS). https://www.pwc.com/gx/en/industrial-manufacturing/pdf/intro-implementing-ai-manufacturing.pdf
Queiroz MM, Telles R, Bonilla SH (2019) Blockchain and supply chain management integration: a systematic review of the literature. Supply Chain Manag Int J
Rai et al (2021) Machine learning in manufacturing and industry 4.0 applications. Int J Prod Res 59(16):4773–4778
Rane SB, Potdar PR (2021) Blockchain-IoT-based risk management approach for project procurement process of asset propelled industries. Int J Procure Manag 14(5)
Rane SB, Thakker SV (2020) Green procurement process model based on blockchain-IoT integrated architecture for a sustainable business. Manag Environ Qual 31(3):741–763
Rejeb A, Keogh JG, Treiblmaier H (2019) Leveraging the internet of things and blockchain technology in supply chain management. Fut Internet 11(7):161
Rejeb A, Keogh JG, Simske SJ, Stafford T, Treiblmaier H (2021) Potentials of blockchain technologies for supply chain collaboration: a conceptual framework. Int J Logist Manag

Roberson CM (2019) Preparing for the unknown in your supply chain. Forbes. https://www.forbes.com/sites/cathymorrowroberson/2019/10/07/preparing-for-the-unknown-in-your-supply-chain/?sh=6e43b68839ff. Accessed 4 Sep 2021
Rodrigo MNN, Perera S, Senaratne S, Jin X (2018) Blockchain for construction supply chains: a literature synthesis. In: Proceedings of ICEC-PAQS conference 2018, Sydney, Australia
Rogers E (1995) Diffusion of innovations, 4th edn. Free Press, New York, NY
Rogerson M, Parry GC (2020) Blockchain: case studies in food supply chain visibility. Supply Chain Manag Int J
Rohaan D, Topan E, Groothuis-Oudshoorn CGM (2022) Using supervised machine learning for B2B sales forecasting: a case study of spare parts sales forecasting at an after-sales service provider. Expert Syst Appl 188
Rymarczyk T, Kłosowski G (2018) The use of artificial intelligence in automated inhouse logistics centres. IAPGOŚ 1(2018):48–51
Saberi S, Kouhizadeh M, Sarkis J, Shen L (2019) Blockchain technology and its relationships to sustainable supply chain management. Int J Prod Res 57(7):2117–2135
Santhi RA, Muthuswamy P (2022) Influence of blockchain technology in manufacturing supply chain and logistics. Logistics 6(1):15
Sarpatwar K, Vaculin R, Min H, Su G, Heath T, Ganapavarapu G, Dillenberger D (2019) Towards enabling trusted artificial intelligence via blockchain. In: Policy-based autonomic data governance. Springer, pp 137–153. https://doi.org/10.1007/978-3-030-17277-0_8
Schulze-Horn et al (2020) Artificial intelligence in purchasing: facilitating mechanism design-based negotiations. Appl Artif Intell 34(8):618–642
Shao D, Kombe C, Saxena S (2022) An ensemble design of a cash crops-warehouse receipt system (WRS) based on blockchain smart contracts. J Agribus Dev Emerg Econ
Sharma M, Joshi S, Luthra S, Kumar A (2021) Managing disruptions and risks amidst COVID-19 outbreaks: role of blockchain technology in developing resilient food supply chains. Oper Manag Res 1–14
Sheffi Y (2015) Preparing for disruptions through early detection. MIT Sloan Manag Rev 57(1):31
Shwetha AN, Prabodh CP (2021) A comprehensive review of blockchain based solutions in food supply chain management. In: 2021 5th international conference on computing methodologies and communication (ICCMC), pp 519–525
Singh S, Kumar R, Panchal R, Tiwari MK (2021) Impact of COVID-19 on logistics systems and disruptions in food supply chain. Int J Prod Res 59(7):1993–2008
Sohrabpour et al (2021c) Export sales forecasting using artificial intelligence. Technol Forecast Soc Change 163
Šustrová T (2016) A suitable artificial intelligence model for inventory level optimization. Trends Econ Manag 10(25)
Swanson EB (2004) How is an it innovation assimilated. In: IFIP working conference on IT innovation for adaptability and competitiveness. Springer, Boston, MA, pp 267–287
Taboada I, Shee H (2021) Understanding 5G technology for future supply chain management. Int J Log Res Appl 24(4):392–406
Tang CS (2006) Perspectives in supply chain risk management. Int J Prod Econ 103(2):451–488
Tanwar S, Bhatia Q, Patel P, Kumari A, Singh PK, Hong W-C (2019) Machine learning adoption in blockchain-based smart applications: the challenges, and a way forward. IEEE Access 8:474–488. https://doi.org/10.1109/ACCESS.2019.2961372
Thamer H, Börold A, Benggolo A, Freitag M (2018) Artificial intelligence in warehouse automation for flexible material handling. In: Conference: 9th international scientific symposium on logistics, Magdeburg, Germany
Thomas KJ (2020) Supply chain management and dealing with the unknown. https://www.worldlocity.com/post/supply-chain-management-and-dealing-with-the-unknown. Accessed 10 July 2021
Toorajipour et al (2021) Artificial intelligence in supply chain management: a systematic literature review. J Bus Res 122:502–517

Torky M, Hassanein AE (2020) Integrating blockchain and the internet of things in precision agriculture: analysis, opportunities, and challenges. Comput Electron Agricul 105476
Tornatzky LG, Fleischer M (1990) The processes of technological innovation. Lexington Books, Lexington, MA
Treiblmaier H (2018) The impact of the blockchain on the supply chain: a theory-based research framework and a call for action. Supply Chain Manag Int J
Tzachor A (2020) Artificial intelligence for agricultural supply chain risk management: Preliminary prioritizations and constraints for the deployment of AI in food chains assessed by CGIAR scientists. In: CGIAR Big Data Platform, 26 p
UN News (2022) Ukraine: 'We need peace now' declares Guterres, warning of global hunger meltdown. United Nations. https://news.un.org/en/story/2022/03/1113882. Accessed 15 Mar 2022
Wang L (2019) From intelligence science to intelligent manufacturing. Engineering 5(4):615–618
Wang S (2021) Artificial intelligence applications in the new model of logistics development based on wireless communication technology. Sci Programm
Wang S, Qu X (2019) Blockchain applications in shipping, transportation, logistics, and supply chain. In: Qu X, Zhen L, Howlett R, Jain L (eds) Smart transportation systems 2019. Smart innovation, systems and technologies, vol 149, pp 225–231. Springer
Wang J, Liu J, Zheng L (2018) Construction of VMI mode supply chain management system based on block chain. In: Vaidya J, Li J (eds) Algorithms and architectures for parallel processing. ICA3PP 2018. Lecture Notes in Computer Science. Springer, p 11337
Wang Z, Wang T, Hu H, Gong J, Ren X, Xiao Q (2020) Blockchain-based framework for improving supply chain traceability and information sharing in precast construction. Autom Constr 111:103063
Wang Y, Men S, Guo T (2021) Application of blockchain technology in value chain of procurement in manufacturing enterprises. Wirel Commun Mob Comput
Wei WQ, Li LY (2022) The impact of artificial intelligence on the mental health of manufacturing workers: the mediating role of overtime work and the work environment. Front Publ Health 10
Wharton Business Daily (2020) Coronavirus and supply chain disruption: what firms can learn. https://knowledge.wharton.upenn.edu/article/veeraraghavan-supply-chain/. Accessed 31 Aug 2021
Wu T, Blackhurst J, O'grady P (2007) Methodology for supply chain disruption analysis. Int J Prod Res 45(7):1665–1682
Xu S, Zhang X, Feng L, Yang W (2020) Disruption risks in supply chain management: a literature review based on bibliometric analysis. Int J Prod Res 58(11):3508–3526
Yadav VS, Singh AR, Raut RD, Govindarajan UH (2020) Blockchain technology adoption barriers in the Indian agricultural supply chain: an integrated approach. Resour Conserv Recycl 161:104877
Yang B, Huang Z (2021) The application research of artificial intelligence in logistics field. In: 2021 international conference on intelligent computing, automation and applications (ICAA)
Yang JX, Li L, Rasul M (2021) Warehouse management models using artificial intelligence technology with application at receiving stage—A review. Int J Mach Learn Comput 11(3):242–249
Younis H, Sundarakani B, Alsharairi M (2021) Applications of artificial intelligence and machine learning within supply chains: systematic review and future research directions. J Modell Manag
Yuan H, Qiu H, Bi Y, Chang SH, Lam A (2019) Analysis of coordination mechanism of supply chain management information system from the perspective of block chain. Inf Syst e-Bus Manag 1–23
Zhang Y (2019) The application of artificial intelligence in logistics and express delivery. J Phys: Conf Ser 1325:012085
Zhang et al (2020) Artificial Intelligence in advanced manufacturing: current status and future outlook. J Manuf Sci Eng 142:1–53
Zuo Y (2021) Making smart manufacturing smarter—A survey on blockchain technology in industry 4.0. Enterp Inf Syst 15(10)

Supply Chain Resilience Strategies for Times of Unprecedented Uncertainty

Hemendra Nath Roy, Eman Almehdawe, and Golam Kabir

1 Introduction

The movement of necessary goods and services has been a fundamental logistical challenge since the primeval stages of human society, and it has since evolved into a critical branch of contemporary business practice commonly known as "supply chain management." The concept of supply chain management was developed and implemented by the US Army during World War II to mitigate risk and disruption in the management and distribution of supplies to the front lines. One of the key events that motivated the development of supply chain management was Japan's capture of Java—one of the world's main producers of quinine at the time—as the US and the rest of the world did not have an alternative source of this compound (Bechtold et al. 2021). Most recently, the supply chain has been challenged by the unprecedented mass disruption created by the global spread of COVID-19 (Manupati et al. 2022). Furthermore, the recovery of the global supply chain has recently experienced further setbacks and disruptions due to the effects of the ongoing Russia–Ukraine war. Thus, supply chain management is mainly a process of developing strategies and actions for mitigating risk and disruption in anticipation or response to unforeseen disruptive events or periods of uncertainty. Institutional interest in the supply chain has been increasing since the 1980s when businesses and corporations began to search for low-cost sources for each material rather than relying on the existing ones (Lummus and Vokurka 1999). After a decade, the term, "supply chain," entered into common usage in academic research, and it has since become a vital branch of modern research,

H. N. Roy · G. Kabir (✉)
Industrial Systems Engineering, Faculty of Engineering and Applied Science, University of Regina, Regina, SK S4S 0A2, Canada
e-mail: Golam.Kabir@uregina.ca

E. Almehdawe
Faculty of Business Administration, University of Regina, Regina, SK S4S 0A2, Canada

S. K. Paul et al. (eds.), *Supply Chain Risk and Disruption Management*, Flexible Systems Management, https://doi.org/10.1007/978-981-99-2629-9_5

especially given the intrinsic interface between the supply chain and the individual consumer (Alfalla-Luque and Medina-López 2009).

Most recent research has focused on maximizing the supply chain's operational efficiency and resilience against disruption, especially in light of the challenges caused by COVID-19. For example, Paul and Chowdhury (2021) investigated the production recovery strategies of a manufacturing unit during COVID-19. Some research has focused on the development of resilience strategies for managing the impact and recovery from disruptions. While significant attention has been devoted to the food and healthcare industries during the COVID-19 period, researchers have largely overlooked the retail industry (Chowdhury et al. 2021). This oversight is notable, as the retail supply chain (RSC) has emerged as the most public-centric, and any disruption to this network affects the customer base negatively. Retailers are the channel leader of the entire supply chain and are linked either directly or indirectly with its functions. Consequently, they are the first to feel the effects of any disruptions in the chain (Alikhani et al. 2021). Some researchers have explored disruption mitigation strategies that involve changing production and distribution planning in response to any disruption in the RSC (Paul et al. 2017). Although professionals and academic researchers have achieved improvements to the RSC over the past few decades, the substantial global disruption caused by the COVID-19 pandemic has intensified awareness regarding risks to the supply chain due to disruption, and the need for strategies to deal with such events (Bechtold et al. 2021). This is a significant development, as the RSC has always consisted of complex networks existing in a dynamic environment that is highly prone to external and internal uncertainty (Sharma et al. 2021).

Most research has acknowledged the effectiveness of integrating different strategies to improve the resiliency of the RSC. For this reason, it is vital to understand the nature of supply chain dynamics and equip the RSC with an effective set of strategies to tackle any future uncertainty and disruptions to retail businesses. In Sect. 2, we discuss supply chain risks along with disruptions that could potentially affect the RSC. In Sect. 3, we define SCR and review relevant literature. While in Sect. 4, we make a major contribution to building resiliency in the RSC by highlighting strategies for dealing with any potential disruption that may arise. In Sect. 5, we highlight available mathematical models that have been developed and incorporated into several resilient strategies. Finally, in Sect. 6, we summarize our findings and discuss potential directions for future research.

2 Supply Chain Uncertainty, Risk, and Disruptions

Uncertainty has two faces by nature. Some uncertainties can lead to positive outcomes; these uncertainties are commonly known as opportunities. The other face of uncertainty involves negative outcomes, also known as or risks, which may result in damage of any form to resources. We focus on this latter face of uncertainty. Risk can be articulated in terms of two variables: the probability of an uncertain event's occurrence and the gravity of the event's consequences vis-à-vis resources (Cagliano et al. 2012). From a supply chain perspective, risks are always treated as negative factors that may trigger a disturbance in the network. A risk is any hazard that may create a disturbance in any of the positions of the supply network thereby hindering the normal flow of material, products, information, or money. Due to the coexistence of both external and internal dynamic variables, the supply chain is highly susceptible to unprecedented levels of risk, with certain risks contributing more to network disruptions. Identifying risk and anticipating its effects is a very critical aspect of supply chain risk management. However, managing risk throughout the entire supply chain is astronomically more complex than at a unit level (Jüttner 2005). Hence, the main focus in supply chain management is to understand any potential risks and develop strategies for managing these negative uncertainties.

Properly mapping the relevant risks and their consequences is integral to developing effective strategies for addressing potential disorder in any of the elements of the RSC network. In this section, we summarize the possible risks and disruptions that have been highlighted by various researchers in recent years, and we divide them into several categories to gain a better understanding of their orientations and impacts. Risks can be placed into one of two major categories depending on their frequency and impact: Type I risks, which are risks that occur frequently, but have a relatively small impact on the supply chain; or Type II risks, which are risks that arise due to unprecedented or rare events and severely impact the RSC (Alikhani et al. 2021; Tang and Musa 2011). Risks can also be classified based on the affected domains in the supply chain: demand-side risks refer to risks related to the node of customers zones, retail stores, and their supporting DCs, while supply-side risks affect manufacturing units and their storage facilities, as well as suppliers and producers of raw-materials (Alikhani et al. 2021; Jüttner 2005). However, care must be taken when using this classification system, as supply and demand zones may vary depending on the industry.

The impacts of risks can be categorized based on their area of impact; that is, whether the risk impacts the material, financial, or information flow. Depending on the type of risk, the primary impact can be ascribed to one or a combination of these three categories. For instance, a disruption to capacity or the transportation network will negatively affect the flow of materials; disturbances in the information technology (IT) infrastructure can disrupt the flow of information; and events like instant bankruptcy or financial weakness among beneficiaries may disrupt the flow of financial resources (Tang and Musa 2011). However, in most cases, the end impact is financial loss for the organization. Similarly, sources of risk can be classified in various ways. For instance, sources of risk can be divided into three major groups in order to better understand them and develop better approaches to dealing with them. Firstly, environmental sources of risk are not directly part of the supply network; rather, they emanate from the external environment in which the supply function is operating. We can further reclassify this source of risk into two sub-factors: man-made factors, including fuel cost fluctuations, changes in trade agreements, war, and political unrest; and natural factors, such as tornadoes, fires, earthquakes, or pandemics. Secondly, process-based sources of risk related to any of the processes performed in the various nodes to support the supply chain, including manufacturing processes, financial processes, planning processes, and scheduling processes. These sources of risk are significant, as any disturbance in these processes contributes to deviation from the objectives of the supply system. Finally, elements of the supply chain control system, such as the operating policy and procedures, can eventually act as a source of risk, as any of these control mechanisms can come to restrict the flexibility of the whole system and cause a disruption (Jüttner 2005).

Risk can also be classified based on external and internal sources. Internal sources of risk include factors that are regulated by the organization, such as strategy, operational policy, and procedure. In contrast, external sources of risk include all other factors, such as social, political, legal, and other peripheral factors (Cagliano et al. 2012). As businesses have become more exposed globally, external sources of risk have grown in importance. Usually, the partnering organization's management plays a key role in managing risk at the operational and strategic levels, especially with regards to sourcing, procurement, transportation, and third-party logistics (3PL). Table 1 presents a list of risk sources and their respective impacts on the different echelons of the supply chain that have been identified by researchers and practitioners. The impacts are further classified based on the duration of the disruption. Most trigger a short-term disruption (i.e., lasting 1–2 months), whereas few last for longer periods (i.e., >2 months), for example, the ongoing chip shortage that has been adversely affecting automobile industry throughout the COVID-19 pandemic (Wu et al. 2021).

To illustrate the impact on different supply chain functions and elements, we use three categories: Severe (S), Moderate (M), and Neutral (N). Depending on the initial extent of the impact, few risks will affect all networks simultaneously; rather, most will impact the particular node or link locally. We have denoted this as the epicenter of the effect and divided it into two levels based on the instant shock of the disruption (i.e., local or global). While some disruptions might only affect the node, these disruptions may also lead to disturbances upstream or downstream of the supply chain over time. Such disturbances are also classified as local. For example, a natural disaster in a country may destroy all of the nodes in that area instantly, but further impacts on the connected links or nodes may take days, weeks, or months to emerge. In some cases, the risk and the associated disruption can spread across the globe and turn into a global disruption. COVID-19 originally disrupted nodes and links in China where it was first detected, but it would quickly come to impact the global supply commitment for many industries. While some disruption immediately impacts the entire supply network, increases in oil price have an instant global effect on the costs of logistics (i.e., higher oil prices result in higher logistic costs).

The complexity of the RSC is easy to see, as businesses deal with the basic portfolio of consumer goods, including food, clothing, medicine, stationaries, sports, kid's essentials, and living essentials, among others. Thus, the RSC directly impacts the ability of consumers to meet their basic. Hence, the need to bolster the resiliency of the RSC is of paramount importance. As the last echelon of the business retail store is directly concerned with satisfying the end customer's demand, the initial ripple effect of changes in demand emanates from this point in the chain. During

Table 1 Sources of supply chain risks and their impact

#	Sources of risk	Risk	Examples	Impact on supply chain function			Impact on different echelon				Duration		Severity		Epicenter	
				MF	FF	IF	RM	MA	DC	RC	ST	LT	T1	T2	L	G
1	External	Natural disaster (excluding pandemic) (Novoszel and Wakolbinger 2022)	Earthquake, fire, tsunami, etc.	S	M	S/M	S	S/M	M	M	✓	✓	✓	✓	✓	
2	External	Pandemic (Novoszel and Wakolbinger 2022)	COVID-19, foot and mouth outbreak, etc.	S	S/M	M	S/M	S	M	S		✓		✓		✓
3	External	Man-made disaster (Novoszel and Wakolbinger 2022)	Accidents, cyber-attack, etc.	M	S/M	S/M	S/M	S/M	S/M	M	✓		✓		✓	
4	External	Political (polices and diplomatic relations) (Cagliano et al. 2012)	Shifts in trade agreement, etc.	M	M	N	M	M	N/M	N/M		✓		✓		✓
5	External	Economic war (Cagliano et al. 2012)	Fuel price fluctuation, currency fluctuation, etc.	M	M	N	S	S	M	M		✓		✓	✓	✓
6	External	Legal (Cagliano et al. 2012)	Tariff change, change of Environmental bar, etc.	M	M	N	N	M	S	S		✓	✓		✓	
7	External	Industrial context (Cagliano et al. 2012)	Technology change, resource price increase	M	S	N/M	M	S	N	M	✓	✓	✓		✓	
8	External	Partnering organizations (Cagliano et al. 2012)	Material delivery delay, contract breaching, etc.	S	S	M	S	S	M	M	✓			✓	✓	
9	Inter nal	Process, operational/technical (Cagliano et al. 2012; Jüttner 2005)	Manufacturing process, material handling process, etc.	M	M	M	N	M	M	M	✓		✓		✓	
10		Controlling policy (Jüttner 2005)	Inventory policy, operating policy, etc.	S	M	M	N	M	M	M	✓		✓			✓

MF = Material flow	S = Severe	RM = Raw-material supplier	ST = Short-Term	L = Local
FF = Financial flow	M = Moderate	MA = Manufacturer	LT = Long-Term	G = Global
IF = Information flow	N = Neutral	DC = Distribution center	T1= Type I	
		RC = Retail stores and their customer base	T2 = Type II	

the COVID-19 pandemic, the food supply chain experienced a severe disruption due to the perishability in all types of operations and logistic activities. In addition, the sudden increase in demand during the pandemic was driven in large part by the panic buying that took place in the early stages (Hobbs 2020). Altogether, the effect of disruption on both the demand and supply side resulted in disruptions to the overall supply chain. Thus, identifying and understanding the key supply chain risks can help to underscore the necessity of resiliency in supply chain management. This will be explained in greater detail in the following sections.

3 Supply Chain Resilience (SCR)

Researchers and professionals have been searching for an effective solution to restore business operations to full capacity following disruptions or, ideally, to protect business operations from disruption due to internal or external forces. This has led to a significant increase in interest in SCR. A review of the literature reveals the existence of numerous different definitions for supply chain resiliency. For instance, some research focus on anticipating and overcoming disruptions while defining a resilient supply chain, whereas other studies have emphasized the supply chain's ability to recover from any operational/logistical deviations caused by the disruption (Ambulkar et al. 2015).

Therefore, a resilient supply chain should have the capacity to quickly return to its original stage, or a more favorable one, following a disruption (Behzadi et al. 2017). A more widely used definition of SCR emphasizes multiple factors, including: adaptive capacity; preparedness for disruption; response to disruption; fastest possible recovery time; and cost to evolve in the post-disruptive state to provide superior performance compared to the pre-disruptive phase (Tukamuhabwa et al. 2015). Over the past decade, several studies have focused on developing supply chain resiliency in different types of businesses using both mathematical and theoretical models. In recent years, and especially in light of the COVID-19 pandemic, the term, "resilience," has received significant attention from all types of government and non-government researchers, as well as practitioners. Thus, the main objective of this chapter is to provide an overview of different strategies for increasing supply chain more resilience and mitigating the impacts of disruption. In the following sections, we discuss viable strategies using both theoretical and mathematical resiliency models that have recently been published in the literature.

4 SCR Strategies

Approaches that enhance the supply chain's responsiveness to disruptions and its recovery, thereafter, are known as "resilience strategies." However, resilience strategies are not only intended to function after a disruption; rather, they are also designed to work during normal operating condition as well. Due to the integration of several complex nodes and links, the supply chain always faces uncertainty, and deviations from the standard operating process are very common. Resilience strategies are also effective at responding to these deviations. Over the last 20 years, resilience strategies have been a key topic within the research industry, but the COVID 19 has highlighted the critical nature of this topic on a global scale. As a component of dynamic systems, SCR can be achieved in relation to three domains: engineering, ecological, or evolutionary resilience (Adobor and McMullen 2018).

According to Adobor and McMullen (2018), engineering resilience aims to enhance the supply chain's ability to bounce back to equilibrium after disruptions, which can be achieved through the use of one of five approaches. The first approach is to develop and regularly update a contingency plan to ensure there is a backup plan in place in case the existing resilience plan fails. The second approach involves adopting a delayed product differentiation (DPD) strategy to manage supply and demand. In this approach, a generic version of the product is created to meet almost all types of demand and is customized upon order confirmation to reduce time to market and avoid additional losses related to demand switching. Although DPD is not suitable for all types of businesses, a similar mechanism could be developed to some extent. The third strategy for achieving engineering resilience is to develop a business continuity plan (BCP). This approach is useful for managing a business during rare disasters, as it involves identifying the critical nodes in the supply chain and assessing how the business would be impacted by their absence. One essential element of this type of system is the presence of a triggering mechanism to warn of disruptions ahead of time (www.thebci.org/). The fourth strategy involves increasing the agility of the supply chain to allow it to adapt to any changes in decision-making during any phase of supply operations. Finally, the fifth strategy is to adopt lean manufacturing concepts like Just in Time (JIT) delivery of products or services, as this can increase supply chain efficiency in post-disruptive situations.

Ecological resilience refers to the degree to which the supply chain is able to adapt to changes in the coexisting elements of the network. To attain this type of resilience, the node and link should have functional redundancy and diversifying capability. Redundancy can be achieved in many ways, such as by maintaining spare capacity, using multiple active capacities, and keeping alternative transport systems. Diversification can be defined as the ability to change the direction of the business to match the situational flow. For example, diversification may entail changing the products offered by a business or changing how the business operates. Finally, evolutionary resilience is a purely post-disruptive strategy that is meant to complement strategies aimed at enhancing engineering and ecological resilience. Evolutionary resilience aims to take the organization to the next level through visionary leadership and implementing the changes required to improve the overall situation in the supply chain. The above three dimensions of resilience can bolster the RSC by ensuring its readiness, ability to respond to and recover from disruptions, and future growth or renewal.

Resilience strategies can be further classified into two major categories based on when they are applied: proactive strategies, which are deployed in the pre-disruption phase; and reactive strategies, which are deployed during disturbance-causing events (Tukamuhabwa et al. 2015). Table 2 provides a summary of proactive and reactive strategies, along with their effectiveness against different types of disruptions. Our literature review focused on identifying the latest mathematical modeling approaches to achieving supply chain resiliency and their uses. In total, we identified 13 strategies that can be used to improve RSC network resilience, while simultaneously minimizing total supply chain costs.

Facility fortification is considered one of the key proactive resilience strategies (Hasani and Khosrojerdi 2016; Mahmoodjanloo et al. 2016). Simply put, facility fortification comprises additional steps that can lower the impact of disruption if occurred—for example, designing the basement of a manufacturing facility to be above flood levels or building additional structures to significantly lower the cost of damage from a flood, cyclone, or any other water-related disaster. Facility fortification generally entails implementing a set of proactive measures, such as installing structural reinforcements, procuring backup equipment, conducting preventive maintenance, and increasing monitoring (Salehi Sadghiani et al. 2015). Fortification can be divided into several levels in each node. For example, a base structure can be fortified to one of three different levels: primary fortification, which entails meeting locally enforced minimum guidelines for the structure in question; maximum fortification, which is achieved by adhering to the maximum safety considerations as detailed by consultation with local experts and global standards; and super fortification, which applies additional safeguards to the maximum fortification level. The super fortification level can also be considered a set-up cost in situations where a new node needs to be set up or an existing node needs to be relocated to a better place. Similarly, in most cases, manufacturers and service providers are primarily focused on their profitability, which leads them to overlook some of the mandatory cost functions that do not contribute directly to the node's reliability. For example, the implementation of proper health and safety policies and procedures may reduce the rate of accidents, which can in turn reduce resource losses by eliminating downtime and employee absenteeism. However, investment in this area is not commonly perceived to have a direct influence on production or service provision; rather, it is perceived as creating long-term indirect benefits in relation to supply chain reliability. Thus, collaborative investment in this area can improve working conditions, thereby reducing the chances of man-made accidents. There are myriad such opportunities for improvement when developing a new vendor or manufacturer, and these efforts can save considerable resources in a long run, even after an unprecedented disruption.

According to the literature, DCs are comparatively less exposed to disruption; however, when partial disruptions occur, it is usually due to a severe disaster (Tofighi et al. 2016). Thus, we can assume that the probability of disruption in a set of DCs is relatively low, and that the product flow will be maintained via alternate channels if the main channel is disrupted. ***Inventory prepositioning*** is another proactive technique that entails the accumulation of safety stock in DCs (Lücker et al. 2019). This proactive measure creates redundancy in the forward supply system by becoming a

Table 2 Supply chain resilience strategies and their effectiveness in tackling different types of disruption

#S	Type of strategies	Strategies	Resiliency on supply chain function			Resiliency on different echelons				Reduction of impact		Reduction of impact		Tackling the effect on epicenter	
			MF	FF	IF	RM	MA	DC	RC	ST	LT	T1	T2	L	G
S1	Proactive	Facility fortification	✓		✓	✓	✓	✓	✓	✓	✓	✓	✓	✓	
S2		Inventory prepositioning	✓					✓	✓	✓		✓		✓	✓
S3		Direct to store delivery (DSD)	✓						✓	✓	✓	✓	✓	✓	
S4		Vendor managed inventory (VMI)	✓						✓	✓	✓	✓	✓	✓	✓
S5		Multiple sets of nodes	✓			✓	✓	✓	✓	✓	✓	✓	✓	✓	✓
S6		Centralized management system	✓	✓	✓	✓	✓	✓	✓	✓	✓	✓	✓	✓	✓
S7		Supply chain digitalization		✓	✓	✓	✓	✓	✓	✓	✓	✓	✓	✓	✓
S8		Reserved capacity	✓	✓		✓	✓	✓	✓	✓	✓	✓	✓	✓	✓
S9	Reactive	Supply chain risk sharing	✓			✓	✓			✓	✓	✓	✓	✓	✓
S10		Inventory sharing	✓					✓	✓	✓	✓	✓	✓	✓	
S11		Changing the product/business	✓	✓			✓	✓	✓	✓	✓	✓	✓	✓	✓
S12		Redesigning supply chain	✓	✓	✓	✓	✓	✓	✓	✓	✓	✓	✓	✓	✓
S13		Agility	✓	✓	✓	✓	✓	✓	✓	✓	✓	✓	✓	✓	✓

MF = Material flow	S = Severe	RM = Raw-material supplier	ST = Short-Term	L = Local
FF = Financial flow	M = Moderate	MA = Manufacturer	LT = Long-Term	G = Global
IF = Information flow	N = Neutral	DC = Distribution center	T1= Type I	
		RC = Retail stores and their customer base	T2 = Type II	

reactive measure during the post-disruption period. However, carrying extra inventory is a cost and a key source of waste in the lean manufacturing concept. For this reason, it is recommended that this tradeoff be assessed, and that surplus inventory be stored to respond to any unexpected disturbances in the network.

Since the success of a retail business relies on customer demand fulfillment, multiple resilience strategies are needed in practice to guarantee that stores have adequate supplies of product. ***Direct to store delivery (DSD)*** is a proactive strategy that is used to enhance confidence in the supply of retail products by bypassing DCs (Tang and Tomlin 2008). In DSD, manufacturers deliver products directly to points of sale/points of consumption from the manufacturing facility or, sometimes, their closest warehouse (Otto et al. 2009). Most perishable items are delivered directly to the store to avoid unnecessary delays at intermediary DCs. However, DSD results in higher transport costs for products that usually ship in quantities less than the truckload. Nevertheless, we consider DSD an alternative delivery configuration that can increase supply chain flexibility.

Vendor-managed inventory (VMI) is another proactive approach that can improve supply chain resiliency. VMI has been used by many retailers since Walmart and Proctor & Gamble first adopted this strategy in 1985. Using the VMI strategy, retailers negotiate special terms and conditions which hold manufacturers liable for inventory on the shelves of retail stores up to a certain limit. In this arrangement, the vendor carries an amount of inventory in a closer warehouse to support retail spontaneously. Furthermore, the vendor is penalized for surplus inventory, with the retailer having a fixed income at the agreed-upon volume, thus preventing loss in the case of short stock (Darwish and Odah 2010). VMI progressively promotes the growth of the vendor's business through proper market research and production scheduling, while simultaneously helping retailers to operate their businesses more efficiently (Achabal et al. 2000). Using the VMI model, retail organizations define the amount of inventory that should be managed through the VMI agreement. This strategy actively increases confidence that retailers will continue to have products during times of uncertainty, while also protecting them against surpluses from falling sales due to changes in demand.

Multiple set covering the immediate successors of the network is another proactive resilience strategy. Multiple set actively reduces the possibility of discontinuation caused by downsizing the number of vulnerable arcs and nodes (Sadghiani et al. 2015). In this strategy, every node in an echelon is covered by multiple sets of its predecessors in a supply chain. One retail outlet should be supplied by multiple DCs so that any disruptions at one DC can be compensated for by the rest of the DCs in the set. The DCs in this set are located in different places with different routes and contexts of operations. Similarly, manufacturers in a set are always located in different geographical areas so that if one country is affected by a disaster, the manufacturing facilities in the other countries can actively satisfy the needs of the affected DCs.

The retail business model mainly consists of retail stores, DCs, and vendors or manufacturers; as such, this model is substantially more complex due to its vertically integrated nature and more tightly coupled in terms of decision-making capabilities

(Bode and Wagner 2015). Tordecilla et al. (2021) have described the supply chain network as a large-scale complex system, noting that its management is both critical and time-sensitive. Sometimes, the supply chain system boundary can extend up to the policy level of government, allowing retailers to negotiate trading terms and conditions in favor of the business centrally (Taleizadeh et al. 2022).

Hence, this dispersed supply chain structure is highly compatible with a centralized management system. This structure allows retailers to take leadership over each echelon and platform during the execution of organizational objectives and easily control nodes that do not directly belong to them, for example, manufacturers or suppliers of raw materials. In this arrangement, there is a centralized relationship, with all of the nodes situated at the end-to-end supply chain of each product. We have formalized this relationship by incorporating an open-costing negotiation method whereby both parties agree to a fixed operational, sales, and governance cost for every product, while also agreeing to open raw-material costs to allow changes in raw material prices to be factored into the product cost transparently. Therefore, our extensive preventative ***centralized management system*** can govern the raw-material price negotiation process to support the listed set of manufacturers, while also allowing centralized control over each node. This strategy will be effective in any situation, as raw materials represent between 60 and 75% of the cost of any product (Asking and Gustavsson 2011), and manufacturers are continually forced to deal with the complexities related to supplier selection and order distribution (Torabi et al. 2015). Therefore, a centralized strategy for managing the raw material supply base is crucial for enhancing resilience throughout the entire supply chain. In addition, a centralized strategy ensures better negotiation on price since the introduction of a large, consolidated volume of materials during negotiation helps to ensure efficient supply planning—and thus, more stable product costs—compared to the more traditional system followed by manufacturers. The effects of disruptions on the manufacturer's end can be mitigated by re-directing the supply of raw materials to the next available manufacturer, thus ensuring that product costs remain stable, with the possible exception of changes in transportation costs due to the change of destination. This centralized strategy enhances supply chain resiliency in three major ways: (i) it keeps the product price at a consistent level with minimum changes; (ii) it maximizes the availability of raw materials and relevant control, thereby increasing the sustainability of businesses in the supply chain; and (iii) it optimizes allocation among vendors. Notably, this management strategy is simultaneously proactive and reactive due to the flexibility afforded by its centralized negotiation process, as well as the ability to implement open costing for all upstream supply nodes that do not belong to the end business organization.

Supply chain digitalization is another important proactive strategy that should be considered in building SCR (Ivanov et al. 2019), as it equips the supply chain to keep pace with Industry 4.0. In this strategy, the entire supply chain network is visualized digitally, and smart decision-making capabilities are enabled via the integration of big data analytics, additive manufacturing, and advanced tracking technology. The integration of this strategy into the retail industry has proven to be highly effective as both a proactive and reactive measure for assessing, dealing with, and recovering from unprecedented disruptions.

Another proactive strategy is ***reserved capacity***, which involves strengthening backup capacity in each node. Reserved capacity is only activated after a disruption has occurred in order to support the continuation of business operations to a limited extent. In this approach, nodes accumulate extra hidden capacity and only utilize it in emergency circumstances to remove bottlenecks (Alikhani et al. 2021). This approach has a cost trade-off and, usually, the selected course of action is to fortify the node's capacity. This strategy is implemented as a proactive initiative while negotiating with manufacturing partners during the supply chain capacity planning phase.

Supply chain risk-sharing is a tactical business initiative to enhance the supply chain's resilience with respect to the cost of damaged resources (Ponomarov and Holcomb 2009). Outsourcing for a particular volume of product or arranging for proper insurance can function as a risk-sharing strategy in the RSC. Product can be lost or damaged during transportation, and such situations pose two major risks that must be addressed: the first is the amount of money lost, and the second is the potential losses in the customer base. In this case, adequate insurance can at least cover the money directly lost from the disruption. Similarly, the insurance on property can recover lost money during the post-disruption period. Supply chain collaboration with a partner company is also an approach to risk-sharing. Special price discounts on pre-agreed contracts and the provision of support from vertically connected partners who are not impacted (or less impacted) by a major disruption can help reduce the overall supply chain cost afterward. Hence, this strategy can work as one of the key reactive strategies in building SCR.

Inventory sharing is a reactive strategy that helps to ensure demand can be met in the post-disruption period (Alikhani et al. 2021). In this strategy, inventory is distributed throughout the same echelon to ensure that the next echelon is supported in a post-disaster situation. This is also known as inventory transshipment wherein an alternatively available network is employed to create post-disaster resiliency and the available network is sued to satisfy customer demand by minimizing total transport cost. This strategy can further be extended by including the transfer of goods/merchandise among retailers, as this helps retailers to satisfy their customers' demand while simultaneously reducing extra inventory kept as safety stock. This allows retailers to ship unsold inventory to other stores to help meet demand. Furthermore, this flexibility can also save transport cost in cases where one retailer receives a full truckload from a DC and then distributes the product among other retail centers within the city.

Changing the product or business type in response to the market situation during or after the disruption event is an example of a very aggressive reactive strategy (Chen et al. 2021). For example, there was a sudden increase in demand for personal protective equipment, sanitizing products, and food during the COVID-19 pandemic. While some retailers experienced significant hardship as a result of lockdowns and restrictions, those whose supply chains were flexible enough to allow them to convert their mainstream product supply into emergency supplies exhibited much greater levels of resiliency. Sometimes, a business' decision to modify its product or marketing channel, thus adding resilience to its business model, may ultimately be what ensures its survival in a post-disruptive environment (Benedek et al. 2021). A slight modification to the product can allow a business to better respond to customer demand and to overcome the disruption. In this approach, there is always an alternative bill of material (BOM) introduced, or the change is executed by keeping a major part of the BOM the same so it can be fitted for any required change in the post-disruptive market (Hasani and Khosrojerdi 2016).

Sometimes the existing supply chain model is susceptible to high exposure in the event of a disruption. Such situations can be improved by ***redesigning the supply chain***, which may require setting up new establishments or phasing out some structures. This can be the most expensive resilience strategy, but it enables trade-offs, so it is generally viewed as a viable last resort option (Blackhurst et al. 2005). In some cases, more expensive modes of transportation can be traded off to enhance agility and safety during the transport of materials. Similarly, changing all defective machines in a series of operations at the same time could ensure the timely shipment of goods to quality-sensitive customers rather than being late every time. During the COVID-19 pandemic and the post-pandemic period, retailers in many industries added or extended their Omni channel sales processes or online delivery to maximize sales (Fortuna et al. 2021). Thus, redesigning the supply chain is a situation-driven reactive strategy for restoring or even improving a business in any post-disruption period.

Finally, ***agility*** is one of the defining features of supply chain resiliency. All of the aforementioned strategies for improving resiliency are ruled by agility, so it is difficult to define it as a separate strategy. However, agility has a significant influence on resilience in the manufacturing supply chain, as it ensures quick adaptability to change (Tarigan et al. 2021). Agility influences the adoptive response in a post-disaster scenario, thus accelerating resiliency. Therefore, in a dynamic RSC, agility plays a vital role by making the overall supply chain resilient.

4.1 Management Approaches in SCR

Supply chain management requires both soft and hard skills. A proper combination of these skills is necessary for making the supply chain more flexible, which is the prerequisite of achieving resiliency in practice. Undoubtedly, a highly flexible and adaptive management structure is the key to achieving ecological resilience, as the supply chain function is highly dynamic. Indeed, it is necessary to have a skilled and knowledgeable team who can deal with change promptly, such as in a situation requiring a product or business change. A strong change management approach is required to coordinate all these strategies and to tackle the disruption. Depending on the operating context, the supply chain management structure may need to be centralized, decentralized, or, in some cases, in the form of a combinational matrix to cope with the changes in strategies. The management system can also have a critical influence on resiliency by promoting risk management culture, which is key in bringing about holistic change in the organization's approach to dealing with uncertain events, particularly via anticipation, assessment, response, and learning processes (Roeschmann 2014). Thus, an effective management structure is crucial to an organization's ability to achieve resilience in practice.

4.2 Ethical Supply Chain Practices to Achieve SCR

Ethical supply chain practice is an inherent element of a resilient supply chain, as all the operations, communications, and relationships among the nodes and links have an ethical dimension. Environmental sustainability has become a key contemporary business concern, and it has impacted the functioning of the supply chain. Whereas the addition of measures to improve environmental concerns may be seen as a concern for the efficiency and cost of the supply chain function, measures to ensure a more sustainable organization may result in fewer disruptions due to legal regulations (Mari et al. 2014). There are many examples of environmental agencies stepping in and disrupting factory production on the grounds that the factory is an environmental hazard. For instance, many factories in China were shut down as part of an initiative to improve air quality in the lead-up to the 2008 Summer Olympic Games. Such actions have had a significant impact on the predecessor and successor nodes of these factories, and, in some cases, the disruption has been long-term. This example demonstrates how adopting a green approach can make an organization a more reliable and stronger node in the supply network.

5 Supply Chain Models with Resilience Strategy

The RSC has become so complex that it is becoming increasingly difficult to manage it using traditional methods. In response to this challenge, researchers have begun to focus on deriving the supply chain function mathematically to explore different approaches. The main objective of this type of research is to convert theoretical resilience strategies into numerical formulas under multiple scenarios to generate precise business insights. In this section, we highlight some studies that have presented mathematical models and identify the latest resilience strategies in the different echelons of the supply chain. While priority is given to research related to the RSC, we also include a few other studies related to supply chain resilience to highlight future research directions in the retail industry. Finally, we summarize the models and their underlying strategies in Table 3 to provide a benchmark of the different techniques that have been used by researchers and to identify the scope of future research.

Table 3 Mathematical model with different resilience strategies

#	Reference of the Model	Echelon of Supply Chain					Resilience Strategies Used												
		Raw-Material Supplier	Manufacturer	DC	Retail Store	Customer Zone	Facility Fortification	Inventory Prepositionin g	Direct to Store Delivery (DSD)	Vendor Managed Inventory (VMI)	Multiple Set of Node	Centralized Management system	Supply Chain Digitalization	Reserved Capacity	Supply Chain Risk Sharing	Inventory Sharing	Changing the Product/Business	Redesigning Supply Chain	Agility
1	Kungwalsong et al. (2022)	✓	✓	✓	✓	✓		✓			✓	✓						✓	
2	Sawik (2022)		✓			✓		✓			✓								✓
3	Taleizadeh et al. (2022)	✓	✓		✓	✓	✓				✓	✓		✓					
4	Manupati et al. (2022)	✓	✓	✓		✓					✓		✓						✓
5	Lotfi et al. (2022)		✓			✓				✓									
6	Alikhani et al. (2021)		✓	✓	✓	✓	✓	✓	✓		✓			✓		✓			
7	Shao (2021)		✓		✓	✓			✓		✓		✓					✓	
8	Chen et al. (2021)	✓	✓		✓	✓					✓						✓	✓	
9	Hasani et al. (2021)		✓	✓	✓	✓	✓	✓			✓		✓			✓			
10	Fattahi et al. (2020)		✓	✓		✓	✓	✓											✓
11	Snoeck et al. (2019)		✓			✓		✓			✓					✓		✓	
12	Fattahi et al. (2018)		✓	✓		✓	✓	✓			✓							✓	
13	Ye et al. (2018)	✓	✓		✓	✓									✓				
14	Fattahi et al. (2017)		✓	✓	✓	✓	✓	✓			✓					✓			✓
15	Lücker and Seifert (2017)		✓			✓		✓			✓								✓
16	Lim et al. (2017)			✓		✓		✓			✓					✓			
17	Rezapour et al. (2017)	✓	✓		✓	✓			✓		✓								
18	Mohammaddust et al. (2017)		✓	✓	✓	✓		✓			✓			✓		✓			
19	Jabbarzadeh et al. (2016)		✓			✓	✓				✓								
20	Hasani and Khosrojerdi (2016)	✓	✓	✓		✓	✓	✓			✓						✓		
21	Yan and Zhao (2015)		✓		✓	✓			✓							✓			
22	Salehi Sadghiani et al. (2015)		✓		✓	✓					✓					✓			
23	Losada et al. (2012)				✓	✓	✓												✓
24	Wakolbinger and Cruz (2011)		✓		✓	✓									✓				

5.1 Appraisal of SCR

Resiliency can make any business model more sustainable, but there is always a tradeoff between resiliency and investment or expenses (Spiegler et al. 2012). As such, measuring a supply chain's degree of resilience has become a key indicator of its performance. The qualitative approach to assessing supply chain resiliency was common in the literature until the development of a simulation-based quantitative approach that focused on three determinants of the supply network: density, complexity, and node criticality (Falasca et al. 2008). In this chapter, we consider both the qualitative and quantitative approaches to measuring resiliency.

Synchronized measurement of readiness, response, and recovery has been achieved via the use of Integral of the Time Absolute Error (ITAE), a control engineering application that enables comparisons of make-to-stock and make-to-order businesses. Similarly, the three dimensions of the supply chain—namely, absorptive capacity, adaptive capacity, and restorative capacity—have been quantified using the Bayesian network approach (Hosseini et al. 2020). Over the decades, researchers have proposed many resilience strategies, which has made it more complicated to measure the relative strength of each strategy, despite having unique features in each supply chain. For instance, Kumar and Anbanandam (2019) developed a three-phased framework based on the integrated Delphi-Fuzzy method to classify strategies and measure and improve initiatives in attempt to make their model more complete with respect to supply chain dynamics. Elsewhere, Soni et al. (2014) used graph theory to determine the SCR index by assessing the available resilient strategies in the respective model.

In real life, the supply chain is always dynamic in nature as a result of multidimensional and multi-tier metrics; thus, the measurement of supply chain strength in pre-disruptive and post-disruptive situations will be highly dependent on individual nodes and links. While there is relatively little research examining such specific aspects, researchers have developed a three-tiered simulation model that assesses resiliency by aggregating recovery, impact, and performance loss on a time scale (Munoz and Dunbar 2015). A time-series measurement in post-disruptive scenario has been modeled for a two-echelon supply chain (Elluru et al. 2019), but most contemporary supply networks have a minimum of three echelons. Hence, there is a need for a simplified scale to measure the degree of resiliency, and ideally every model should be assessed to identify improvement opportunities for enhancing the sustainability of businesses.

6 Conclusion and Recommendation for Future Research

In this chapter, we make two key contributions to research in the field of supply chain resiliency within the context of retail business organizations. Firstly, we have compiled and discussed a selection of recently published resilience strategies and presented them in a matrix. Specifically, we highlight 13 efficient strategies for managing an uncertain event like the COVID-19. Secondly, we conducted an analysis of the latest mathematical models and further investigated the applied resilience strategies to identify research gaps in the field of RSC resilience models. Most of the surveyed research incorporated strategies such as multiple sets, inventory prepositioning or safety stock, facility fortification, and reserved capacity into their models. Over 90% of the mathematical models used less than 5 of the 13 strategies discussed in this chapter, with only two recent works in the RSC field using more. Indeed, some critical strategies like VMI, centralized management systems, supply chain risk-sharing, and changing products or business models remain largely overlooked by researchers. Our research exposes a remarkable research gap in the formulation of mathematical models using more than five resilience strategies in the RSC field. This gap is significant, as the inclusion of more strategies can further enhance resiliency in the event of unprecedented global disruptions. Additionally, most of the examined models exclude the time to recovery function in multi-product, multi-source, and multi-echelon supply chains, which can be added in future models to map the four parameters of readiness, response, recovery, and future growth or renewal of the supply chain in the time scale. We emphasize this aspect in the measurement approach to supply chain resiliency, as it is highly critical to understanding the effectiveness of any model. Since there is relatively little research focusing on RSC mathematical modeling, the number of studies available for analysis in this literature review was rather limited. Future research on building resiliency in the RSC via mathematical models will enrich the field and enable more resourceful research. Our research attempts to support this future direction by compiling all available resilience strategies with the aim of advancing research efforts focusing on the RSC.

References

Achabal DD, Mcintyre SH, Smith SA, Kalyanam K (2000) A decision support system for vendor managed inventory. J Retail 76(4):430–454. https://doi.org/10.1016/s0022-4359(00)00037-3

Adobor H, McMullen RS (2018) Supply chain resilience: a dynamic and multidimensional approach. Int J Logist Manag 29(4):1451–1471. https://doi.org/10.1108/IJLM-04-2017-0093

Alfalla-Luque R, Medina-López C (2009) Supply chain management: unheard of in the 1970s, core to today's company. Bus Hist 51(2):202–221. https://doi.org/10.1080/00076790902726558

Alikhani R, Torabi SA, Altay N (2021) Retail supply chain network design with concurrent resilience capabilities. Int J Prod Econ 234(2021):108042. https://doi.org/10.1016/j.ijpe.2021.108042

Ambulkar S, Blackhurst J, Jennifer S (2015) Firm's resilience to supply chain disruptions: scale development and empirical examination. J Oper Manag 33–34(2015):111–122. https://doi.org/10.1016/j.jom.2014.11.002

Asking P, Gustavsson S (2011) Cost breakdown analysis—A study of the product costs in the kitchen appliances at IKEA of Sweden [Diploma work no. 56/2011, Chalmers University of Technology]. https://hdl.handle.net/20.500.12380/148459

Bechtold HD, Cruz AT, Kaziny BD (2021) From World War II to COVID-19: a historical perspective on the American medical supply chain. In: Disaster medicine and public health preparedness. Cambridge University Press, pp 1–2. https://doi.org/10.1017/dmp.2021.94

Behzadi G, O'Sullivan MJ, Olsen TL, Scrimgeour F, Zhang A (2017) Robust and resilient strategies for managing supply disruptions in an agribusiness supply chain. Int J Prod Econ 191:207–220. https://doi.org/10.1016/j.ijpe.2017.06.018

Benedek Z, Fertö I, Marreiros CG, de Aguiar PM, Pocol CB, Čechura L, Poder A, Pääso P, Bakucs Z (2021) Farm diversification as a potential success factor for small-scale farmers constrained by COVID-related lockdown. Contributions from a survey conducted in four European countries during the first wave of COVID-19. PLoS One 16(5). https://doi.org/10.1371/journal.pone.0251715

Bode C, Wagner SM (2015) Structural drivers of upstream supply chain complexity and the frequency of supply chain disruptions. J Oper Manag 36:215–228. https://doi.org/10.1016/j.jom.2014.12.004

Cagliano AC, de Marco A, Grimaldi S, Rafele C (2012) An integrated approach to supply chain risk analysis. J Risk Res 15(7):817–840. https://doi.org/10.1080/13669877.2012.666757

Chen J, Wang H, Zhong RY (2021) A supply chain disruption recovery strategy considering product change under COVID-19. J Manuf Syst 60(2021):920–927. https://doi.org/10.1016/j.jmsy.2021.04.004

Chowdhury P, Paul SK, Kaisar S, Moktadir MA (2021) COVID-19 pandemic related supply chain studies: a systematic review. Transp Res Part E: Logist Transp Rev 148(2021):102271. https://doi.org/10.1016/j.tre.2021.102271

Darwish MA, Odah OM (2010) Vendor managed inventory model for single-vendor multi-retailer supply chains. Eur J Oper Res 204(3):473–484. https://doi.org/10.1016/j.ejor.2009.11.023

Elluru S, Gupta H, Kaur H, Singh SP (2019) Proactive and reactive models for disaster resilient supply chain. Ann Oper Res 283(1–2):199–224. https://doi.org/10.1007/s10479-017-2681-2

Falasca M, Zobel CW, Cook D (2008) A decision support framework to assess supply chain resilience. In: Proceedings of the 5th international ISCRAM conference, pp 596–605

Fattahi M, Govindan K, Keyvanshokooh E (2017) Responsive and resilient supply chain network design under operational and disruption risks with delivery lead-time sensitive customers. Transp Res Part E: Logist Transp Rev 101:176–200. https://doi.org/10.1016/j.tre.2017.02.004

Fattahi M, Govindan K, Keyvanshokooh E (2018) A multi-stage stochastic program for supply chain network redesign problem with price-dependent uncertain demands. Comput Oper Res 100:314–332. https://doi.org/10.1016/j.cor.2017.12.016

Fattahi M, Govindan K, Maihami R (2020) Stochastic optimization of disruption-driven supply chain network design with a new resilience metric. Int J Prod Econ 230(2020):107755. https://doi.org/10.1016/j.ijpe.2020.107755

Fortuna F, Risso M, Musso F (2021) Omnichannelling and the predominance of big retailers in the post-covid era. Symphonya. Emerg Issues Manag 2:142–157. https://doi.org/10.4468/2021.2.11fortuna.risso.musso

Hasani A, Khosrojerdi A (2016) Robust global supply chain network design under disruption and uncertainty considering resilience strategies: a parallel memetic algorithm for a real-life case study. Transp Res Part E: Logist Transp Rev 87(2016):20–52. https://doi.org/10.1016/j.tre.2015.12.009

Hasani SMR, Nasiri MM, Torabi SA, Mohtashami Z (2021) A resilient supply chain network for an online retailer: a three-phase robust framework and a case study. Int J Sci Technol. Articles in Press. https://doi.org/10.24200/SCI.2021.55787.4405

Hobbs JE (2020) Food supply chains during the COVID-19 pandemic. Can J Agric Econ 68(2):171–176. https://doi.org/10.1111/cjag.12237

Hosseini S, Ivanov D, Blackhurst J (2020) Conceptualization and measurement of supply chain resilience in an open-system context. IEEE Trans Eng Manag. https://doi.org/10.1109/TEM.2020.3026465

Ivanov D, Dolgui A, Sokolov B (2019) The impact of digital technology and industry 4.0 on the ripple effect and supply chain risk analytics. Int J Prod Res 57(3):829–846. https://doi.org/10.1080/00207543.2018.1488086

Jabbarzadeh A, Fahimnia B, Sheu JB, Moghadam HS (2016) Designing a supply chain resilient to major disruptions and supply/demand interruptions. Transp Res Part B: Methodol 94(2016):121–149. https://doi.org/10.1016/j.trb.2016.09.004

Jüttner U (2005) Supply chain risk management: understanding the business requirements from a practitioner perspective. Int J Logist Manag 16(1):120–141. https://doi.org/10.1108/09574090510617385

Kumar S, Anbanandam R (2019) An integrated Delphi—Fuzzy logic approach for measuring supply chain resilience: an illustrative case from manufacturing industry. Meas Bus Excell 23(3):350–375. https://doi.org/10.1108/MBE-01-2019-0001

Kungwalsong K, Mendoza A, Kamath V, Pazhani S, Marmolejo-Saucedo JA (2022) An application of interactive fuzzy optimization model for redesigning supply chain for resilience. Ann Oper Res. https://doi.org/10.1007/s10479-022-04542-5

Lim MK, Mak HY, Shen ZJM (2017) Agility and proximity considerations in supply chain design. Manag Sci 63(4):1026–1041. https://doi.org/10.1287/mnsc.2015.2380

Losada C, Scaparra MP, O'Hanley JR (2012) Optimizing system resilience: A facility protection model with recovery time. Eur J Oper Res 217(3):519–530. https://doi.org/10.1016/j.ejor.2011.09.044

Lotfi R, Kargar B, Rajabzadeh M, Hesabi F, Özceylan E (2022) Hybrid fuzzy and data-driven robust optimization for resilience and sustainable health care supply chain with vendor-managed inventory approach. Int J Fuzzy Syst 24(2022):1216–1231. https://doi.org/10.1007/s40815-021-01209-4

Lücker F, Seifert RW (2017) Building up resilience in a pharmaceutical supply chain through inventory dual sourcing and agility capacity. Omega (United Kingdom) 73(2017):114–124. https://doi.org/10.1016/j.omega.2017.01.001

Lücker F, Seifert RW, Biçer I (2019) Roles of inventory and reserve capacity in mitigating supply chain disruption risk. Int J Prod Res 57(4):1238–1249. https://doi.org/10.1080/00207543.2018.1504173

Lummus RR, Vokurka RJ (1999) Defining supply chain management: a historical perspective and practical guidelines. Ind Manag Data Syst 99(1):11–17. https://doi.org/10.1108/02635579910243851

Mahmoodjanloo M, Parvasi SP, Ramezanian R (2016) A tri-level covering fortification model for facility protection against disturbance in R-interdiction median problem. Comput Ind Eng 102(2016):219–232. https://doi.org/10.1016/j.cie.2016.11.004

Manupati VK, Schoenherr T, Ramkumar M, Panigrahi S, Sharma Y, Mishra P (2022) Recovery strategies for a disrupted supply chain network: leveraging blockchain technology in pre- and post-disruption scenarios. Int J Prod Econ 245(2022):108389. https://doi.org/10.1016/j.ijpe.2021.108389

Mari SI, Lee YH, Memon MS (2014) Sustainable and resilient supply chain network design under disruption risks. Sustainability (Switzerland) 6(10):6666–6686. https://doi.org/10.3390/su6106666

Mohammaddust F, Rezapour S, Farahani RZ, Mofidfar M, Hill A (2017) Developing lean and responsive supply chains: a robust model for alternative risk mitigation strategies in supply chain designs. Int J Prod Econ 183(2017):632–653. https://doi.org/10.1016/j.ijpe.2015.09.012

Munoz A, Dunbar M (2015) On the quantification of operational supply chain resilience. Int J Prod Res 53(22):6736–6751. https://doi.org/10.1080/00207543.2015.1057296

Novoszel L, Wakolbinger T (2022) Meta-analysis of supply chain disruption research. Oper Res Forum 3:10. https://doi.org/10.1007/s43069-021-00118-4

Otto A, Schoppengerd FJ, Shariatmadari R (2009) Success in the consumer products market—Understanding direct store delivery. In: Direct store delivery. Springer Berlin Heidelberg, pp 1–27. https://doi.org/10.1007/978-3-540-77213-2_1

Paul SK, Chowdhury P (2021) A production recovery plan in manufacturing supply chains for a high-demand item during COVID-19. Int J Phys Distrib Logist Manag 51(2):104–125. https://doi.org/10.1108/IJPDLM-04-2020-0127

Paul SK, Sarker R, Essam D (2017) A quantitative model for disruption mitigation in a supply chain. Euro J Oper Res 257(3):881–895. https://doi.org/10.1016/j.ejor.2016.08.035
Ponomarov SY, Holcomb MC (2009) Understanding the concept of supply chain resilience. Int J Logist Manag 20(1):124–143. https://doi.org/10.1108/09574090910954873
Rezapour S, Farahani RZ, Pourakbar M (2017) Resilient supply chain network design under competition: a case study. Eur J Oper Res 259(3):1017–1035. https://doi.org/10.1016/j.ejor.2016.11.041
Roeschmann AZ (2014) Risk culture: what it is and how it affects an insurer's risk management. Risk Manag Insur Rev 17(2):277–296. https://doi.org/10.1111/rmir.12025
Sadghiani SN, Torabi SA, Sahebjamnia N (2015) Retail supply chain network design under operational and disruption risks. Transp Res Part E: Logist Transp Rev 75(2015):95–114. https://doi.org/10.1016/j.tre.2014.12.015
Salehi Sadghiani N, Torabi SA, Sahebjamnia N (2015) Retail supply chain network design under operational and disruption risks. Transp Res Part E: Logist Transp Rev 75(2015):95–114. https://doi.org/10.1016/j.tre.2014.12.015
Sawik T (2022) Stochastic optimization of supply chain resilience under ripple effect: a COVID-19 pandemic related study. Omega (United Kingdom) 109(2022):102596. https://doi.org/10.1016/j.omega.2022.102596
Shao X (2021) Omnichannel retail move in a dual-channel supply chain. Eur J Oper Res 294(3):936–950. https://doi.org/10.1016/j.ejor.2020.12.008
Sharma M, Luthra S, Joshi S, Kumar A (2021) Accelerating retail supply chain performance against pandemic disruption: adopting resilient strategies to mitigate the long-term effects. J Enterp Inf Manag 34(6):1844–1873. https://doi.org/10.1108/JEIM-07-2020-0286
Snoeck A, Udenio M, Fransoo JC (2019) A stochastic program to evaluate disruption mitigation investments in the supply chain. Eur J Oper Res 274(2):516–530. https://doi.org/10.1016/j.ejor.2018.10.005
Soni U, Jain V, Kumar S (2014) Measuring supply chain resilience using a deterministic modeling approach. Comput Ind Eng 74(2014):11–25. https://doi.org/10.1016/j.cie.2014.04.019
Spiegler VLM, Naim MM, Wikner J (2012) A control engineering approach to the assessment of supply chain resilience. Int J Prod Res 50(21):6162–6187. https://doi.org/10.1080/00207543.2012.710764
Taleizadeh AA, Ahmadzadeh K, Sarker BR, Ghavamifar A (2022) Designing an optimal sustainable supply chain system considering pricing decisions and resilience factors. J Clean Prod 332(2022):129895. https://doi.org/10.1016/j.jclepro.2021.129895
Tang C, Tomlin B (2008) The power of flexibility for mitigating supply chain risks. Int J Prod Econ 116(1):12–27. https://doi.org/10.1016/j.ijpe.2008.07.008
Tang O, Musa SN (2011) Identifying risk issues and research advancements in supply chain risk management. Int J Prod Econ 133(1):25–34. https://doi.org/10.1016/j.ijpe.2010.06.013
Tarigan ZJH, Siagian H, Jie F (2021) Impact of internal integration, supply chain partnership, supply chain agility, and supply chain resilience on sustainable advantage. Sustainability 13(10). https://doi.org/10.3390/su13105460
Tofighi S, Torabi SA, Mansouri SA (2016) Humanitarian logistics network design under mixed uncertainty. Eur J Oper Res 250(1):239–250. https://doi.org/10.1016/j.ejor.2015.08.059
Torabi SA, Baghersad M, Mansouri SA (2015) Resilient supplier selection and order allocation under operational and disruption risks. Transp Res Part E: Logist Transp Rev 79:22–48. https://doi.org/10.1016/j.tre.2015.03.005
Tordecilla RD, Juan AA, Montoya-Torres JR, Quintero-Araujo CL, Panadero J (2021) Simulation-optimization methods for designing and assessing resilient supply chain networks under uncertainty scenarios: a review. In: Simulation modelling practice and theory, vol 106. Elsevier BV. https://doi.org/10.1016/j.simpat.2020.102166
Tukamuhabwa BR, Stevenson M, Busby J, Zorzini M (2015) Supply chain resilience: definition, review and theoretical foundations for further study. In: Int J Prod Res 53(18):5592–5623 (Taylor and Francis Ltd.). https://doi.org/10.1080/00207543.2015.1037934

Wakolbinger T, Cruz JM (2011) Supply chain disruption risk management through strategic information acquisition and sharing and risk-sharing contracts. Int J Prod Res 49(13):4063–4084. https://doi.org/10.1080/00207543.2010.501550

Wu X, Xiling C, Du W (2021) An analysis on the crisis of "chips shortage" in automobile industry. J Phys: Conf Ser 1971(2021):012100. https://doi.org/10.1088/1742-6596/1971/1/012100

Yan X, Zhao H (2015) Inventory sharing and coordination among *n* independent retailers. Eur J Oper Res 243(2):576–587. https://doi.org/10.1016/j.ejor.2014.12.033

Ye F, Hou G, Li Y, Fu S (2018) Managing bioethanol supply chain resiliency: a risk-sharing model to mitigate yield uncertainty risk. Ind Manag Data Syst 118(7):1510–1527. https://doi.org/10.1108/IMDS-09-2017-0429

The Role of Blockchain in Developing Supply Chain Resilience against Disruptions

Hajar SadeghZadeh, Amir Hossein Ansaripoor, and Richard Oloruntoba

1 Introduction and Background

The fragility and lack of resilience of global supply chains have attracted significant and ongoing attention following the outbreak of COVID-19. The persistent global supply chain crisis has been well covered by the media and by scholars. There have been significant delays in the delivery of goods, and rising costs have been the norm in the last year or two. The pandemic-induced global supply chain crisis has imposed a significant additional strain on customer products, manufacturers, carriers, and suppliers worldwide (Sherman 2020; Bailey 2020). This is not the first time that an unexpected disruption such as a disaster has revealed weaknesses in global supply chains. The 2008 global financial crisis (GFC) had significant impacts on global supply chains. The more recent 2011 tidal wave that affected Fukushima, Japan, caused the world's worst nuclear accident which disrupted the automotive industry because 60% of essential auto components were manufactured in the Fukushima area. The Icelandic volcano, Eyjafjallajökull, threw a thick ash cloud across Northern Europe and the European airspace had to be shut down, with airlines unable to fly and air freight disrupted (Langmann 2010). As a result, the manufacture and distribution of many components ceased, and Nissan and BMW quickly halted all production (Theguardian.com 2020). Similarly, the SARS epidemic, which emerged in 2003, affected sections of the electronics supply chain (David 2003). The COVID-19 pandemic has again exposed the risks inherent in global supply chains

H. SadeghZadeh
University of Melbourne, Melbourne, Australia
e-mail: hajar.sadeghzadeh@student.unimelb.edu.au

A. H. Ansaripoor (✉) · R. Oloruntoba
Curtin University, Perth, Australia
e-mail: amir.ansaripoor@curtin.edu.au

R. Oloruntoba
e-mail: richard.oloruntoba@curtin.edu.au

S. K. Paul et al. (eds.), *Supply Chain Risk and Disruption Management*, Flexible Systems Management, https://doi.org/10.1007/978-981-99-2629-9_6

on a much larger scale. Companies worldwide rely on supplies coming from over 12,000 factories located in China, South Korea, and Southeast Asian areas significantly impacted by COVID-19 mitigation measures (Linton 2020).

COVID-19 and the attendant mitigation measures such as border closures and social distancing have resulted in staff shortages and capacity reductions in many industrial sectors, which has made it difficult for managers and industry leaders to react and respond quickly to supply chain disruptions (Butt 2021a, b; Meier and Pinto 2020). Moreover, transportation and logistics have been seriously impacted by such mitigation measures. For instance, the fear of COVID-19 contagion kept many drivers off the road, reducing a large portion of trucking capacity. Overall, COVID-19 has caused delays in deliveries, and entire supply chains have been adversely affected (Chen 2011). Before the COVID-19 pandemic, supply chains were excessively lean (Rivera 2007). They were often built to be low cost and based on Just-in-Time principles (Taylor 1999; Cox 2005). They were also fragmented, long, complex, and often opaque structures. When circumstances were good and stable, a lean supply chain benefited all stakeholders. However, even then, traceability, flexibility, and visibility were limited. However, when the pandemic broke out, underlying problems were exacerbated, and the cumulative consequences resulted in global chaos. Therefore, the pandemic has exposed inherent weaknesses in the design and operation of global supply chains while providing an opportunity to re-think strategic supply chain design to ensure less vulnerability, improved resilience and robustness, and a supply chain that is capable of providing better service to consumers (Boyes 2015).

1.1 Digitalization, Blockchain, and Supply Chain Resilience

The awareness and knowledge of supply chain vulnerability, supply chain resilience, and supply chain risk management (SCRM) have increased over the decade (Peck 2005, 2006; Glickman 2006; Christopher 2004; Pettit 2010), and many scholars have argued for, and advocated the value of, supply chain digitalization and industry 4.0 technologies as the means of mitigating supply chain risks and disruption (e.g., Ivanov 2021; Spieske 2021; Sawik 2022a, b). The authors take the view that, with the appearance of digitization and information analytics, organizations now have the technological tools to offer end-to-end risk management solutions to strengthen supply chain resilience. As some have argued, supply chain digitalization and Industry 4.0 can provide organizations with advanced record-keeping, connections, machines, and integration procedures, thereby improving the cooperation and reliability of members of the supply chain (Raab 2011).

Although these new technologies have a positive impact on productivity and work practices, they can also give rise to malicious electronic cyber-hack or ransomware operations that threaten supply chain security and product safety (Deloitte 2019). In order to resolve these issues, managers and scholars have undertaken initiatives using blockchain technology to secure data, and to ensure system integrity, transparency, security, and protection (Aceto 2019; Kshetri 2017).

Because its benefits need to be completely realized and demonstrated by numerous success stories, it will take some time for blockchain to be acknowledged as a practical tool that can improve security and reduce risk in the supply chain. Blockchain technology allows users to transfer their assets (including intangibles) without the risk of being hacked or the inconvenience of having to create a separate vault that limits the interactions between exchange partners. To expedite the application of blockchain technology, we need to know how blockchain works on a regular basis. Despite its potential benefits, blockchain technology can create a number of challenges when it comes to implementation, such as lack of organizational readiness or technical/infrastructure expertise, issues with scalability, and limited financial resources to invest in blockchain. Therefore, there is an urgent need to develop management strategies that can assist companies to overcome these challenges so that they can fully reap the benefits of blockchain technology. Unfortunately, many blockchain investigations concentrate on bitcoin frameworks, while others are anecdotal studies of potential blockchain applications such as counting smart contracts, financial services, circular economy, and authorizing with limited articles in supply chain management (Nandi 2021; Crosby 2016; Mainelli 2023; Raval 2016; Tapscott 2016; Underwood 2016; Yli-Huumo 2016). Given these gaps in blockchain research, in this chapter, we propose a blockchain architecture to overcome supply chain risks during the Covid-19 pandemic and during the subsequent recovery. Moreover, we suggest several ways by which managers can deploy blockchain to enhance supply chain resilience from a security perspective. We also identify and investigate relevant factors that may influence a company's decision to implement blockchain technology.

While some researchers have explored the application of blockchain in the supply chain, the contribution of this chapter goes beyond this by explaining how blockchain technologies can help manage and predict disruptions and bring greater resilience and balance to the supply chain (Swan 2015a; Ghadge 2019; Taylor 2019; Strozzi 2017).

The rest of this chapter is organized as follows. In Sect. 2, the basic definition and concept of blockchain technology and its general application is discussed. Section 3 explores the application of blockchain technology in non-supply chain management contexts. In Sect. 4, we discuss how blockchain technology can strengthen supply chain resilience. In Sect. 5, we examine the value and role of blockchain in mitigating disruptions and their destructive effects, with the aim of strengthening supply chain resilience. In Sect. 6, we discuss the potential of blockchain to mitigate supply chain disruptions resulting from the pandemic. Section 7 concludes the chapter with a summary and suggestions for further research.

2 Blockchain Technology

Blockchain is a technology that consists of a distributed ledger that stores and exchanges transaction records or other data in a cryptographically secure, decentralized, and immutable format (Swan 2015b). In a blockchain, a 'block' is an information structure that contains a series of data records. Blockchain was created in 2008 by Satoshi Nakamoto, the creator of the Bitcoin whitepaper. Essentially, a sequence of data within a blockchain is an electronically distributed list of data stored by a client or participant on a computer network (Lu 2019). Moreover, blockchain deploys cryptography for transaction processes and verification on the ledger. An important benefit of this decentralized structure is that a business enterprise is not under the control of a single organization. Another benefit is that it can resolve issues of disclosure and accountability between people and institutions whose activities are not strongly regulated (Lu 2019). In blockchain, critical records can be updated securely and accurately in real time, thereby eliminating the need for technologies that are error-prone, which can compromise data confidentiality and security (Casey 2017). Blockchain gives greater transparency and a timely view of activity occurring within the network. Blockchain is attracting interest from supply chain management researchers due to its benefits such as its capability to securely store significant amounts of data, which offers great advantages to individual businesses and to supply chains in general (Kache and Seuring 2017). Also, blockchain enables the encryption and encoding of data, thereby greatly improving data exchange, transparency, performance, and reliability of a network (Misra 2018). The blockchain has four essential traits or attributes (Pattison 2017). First, it is designed to be distributed and it offers synchronization of structures. Hence, corporations can share data with trusted parties in inter-organizational commercial enterprise systems including supply chains and business consortia. Second, blockchain is made up of smart contracts. This is a pre-contract between contributors, and contracts are saved individually inside the blockchain. Smart contracts are computer protocols intended to digitally facilitate, validate, or establish negotiated terms and conditions. Smart contracts enable dependable transactions to occur without the need for third-party intermediaries since all transaction processes are undertaken automatically. Established protocols can help parties decide whether or not a specific transaction, including a particular fee, is authorized (Pilkington 2016). In addition, smart contracts can enable asset verification in a series of transactions involving non-monetary elements (Reyna 2018). This gives the various members of the supply chain network confidence that everyone is following the rules. Third, blockchains are built using Peer-to-Peer (P2P) networks and require all parties to agree to the validity of a transaction. This eliminates faulty or potentially risky transactions from the blockchain database. Fourth, the continuity of the date ensures that the agreed transaction is timely and remains unchanged. This continuity offers a record of the transactions and shows where the asset is located, when it is located, and what is happening to it. There are both public and private blockchains available. The main difference between them is membership, i.e., the users who can become part of the network. Public blockchains are entirely open. In other words, everyone can participate and become part of the network. Networks regularly have incentive mechanisms to encourage more individuals to take part. Bitcoin is one of the most extensive non-public blockchain networks ever created. One disadvantage of private blockchains is the large amount of computing power required to maintain a huge distributed ledger. To obtain consensus, all nodes within the network must be synchronized. This is done by solving a complicated, resource-intensive problem referred to as Proof-of-Work (POW) (Angrish 2018). Furthermore, blockchains require that potential participants be invited to join the network, and participants need to be confirmed. Firms and supply chains typically construct public networks instead of non-public open networks to ensure that access is exclusive. Current participants can determine future entries based totally on hard and fast policies established using the network initiator. Pilkington (2016) Regulators can provide licenses for participation. Alternatively, since the blockchain is decentralized, a consortium may want to decide whether an entity joins the community. Overall, although blockchain technology is relatively new, it has the potential to overcome or mitigate risks associated with the management of supply chains. "Blockchain is a new organizational paradigm for coming across, evaluating, and transferring any quantity (discrete units) of something of value and coordinating all human tasks on a far larger scale than ever before" (Swan 2015b).

3 Applications of Blockchain Technology in Non-SCM Contexts

The commerce and finance sectors are currently making the most use of blockchain technology. Blockchain was specifically designed for Bitcoin, a P2P decentralized virtual cryptocurrency network. Bitcoin has a multibillion dollar international market where buyers can shop anonymously without the control or intervention of authorities. Therefore, it is necessary to address the scope of regulatory problems related to governments and financial institutions in regard to trading with or without bank intervention. All transactions are dispatched to all nodes within the Bitcoin network, and recorded in the general public ledger (Crosby 2016). Blockchain structures are also being used to display monetary institution warranties, follow up monetary exchanges, and mitigate fraud. Using the activated smart contracts, there is network consensus that conditions have been met to allow for automatic payments (Guo 2016). To allow for automatic payments using activated smart contracts network consensus, conditions must be met. Blockchain technology also has the potential to be applied to areas other than coins and currencies, as distributed ledgers can be programmed to record anything of value. This includes birth and death certificates, marriage licenses, documents and deeds, financial reports, scientific procedures, and insurance claims, to name a few (Tapscott 2016). Blockchain applications have also been implemented in healthcare. For example, blockchain is used to track a patient's condition after discharge, and blockchain electronic medical data can be controlled to improve the authentication, privacy, and sharing of records (Armstrong 2018; Marr 2018). In charitable organizations, blockchain has been used to increase the transparency of donations and show the link between donations and project results. In real estate, blockchain is being used to track complex regulatory processes that create value (Marr 2018). Retailers use blockchain to improve monitoring. Monitoring can be implemented in decentralized markets where items and services are traded without intervention (Chakrabarti and Chaudhuri 2017; Grewal 2021). The tourism sector uses blockchain technology to monitor and share information about motel accommodations, and to remove intermediaries for car-sharing and real-time travel (Rejeb 2019; Filimonau 2020). In the media and entertainment sectors, blockchain technologies track intellectual property rights and make payments to artists (Dutra A. 2018). In the public sector area, blockchain is deployed to establish voter identity and increase the credibility of vote counts (Osgood 2016). Overall, blockchain is already helping to improve transparency, transaction speed, and responsiveness. Blockchain is flexible, making it suitable for all trading as it facilitates settlement without dispute. Disputes are eliminated because all the global has a replica of the overall ledger and the invoice is generated automatically.

4 Blockchain Technology for Strengthening Supply Chain Resilience

4.1 Supply Chain Disruptions

Typically, supply chains move products and businesses forward through interactions among trade partners within a complex network. Supply chains are the mechanisms by which goods and services are delivered (Peck 2006). They are multi-level and complex, comprising flows of materials, goods, information, and money which pass within, and between, organizations linked by a range of tangible and intangible facilitators such as processes, relationships, activities, and information systems (Peck 2006). In practice, supply chains are also linked by physical transport, distribution networks, and communication infrastructures often across international borders (Christopher 2000; Peck 2005). However, some of these interactions can increase the vulnerabilities of supply chains which can result in disruptions. To mitigate these vulnerabilities, supply chain experts need to identify the weak links and evaluate the degree of risk posed to the entire supply chain network. Risk evaluations generally entail the following steps:

1. *Identifying the buying and selling partners in the supply chain.* These partners can shape points (or nodes) inside the supply chain network and show the parties who can produce supply chain agreements.
2. *Designing a supply chain or procedure scheme that shows the transaction and its data flow.* These flow depictions (shipment, container, capital, documents, etc.) can reveal potential bottlenecks (i.e., the weakest links) and the level of risk for instance in terms of cyberattacks. Figure 1 depicts the various potential sources of supply chain risk.
3. *Classifying and comparing vulnerabilities.* Risks can be categorized into distinctive classes according to the results of the risk evaluation (e.g., high, medium, low), and the calculated likelihood of the risk occurring. This risk evaluation can then be used to create an action or probability plan to mitigate the risk.

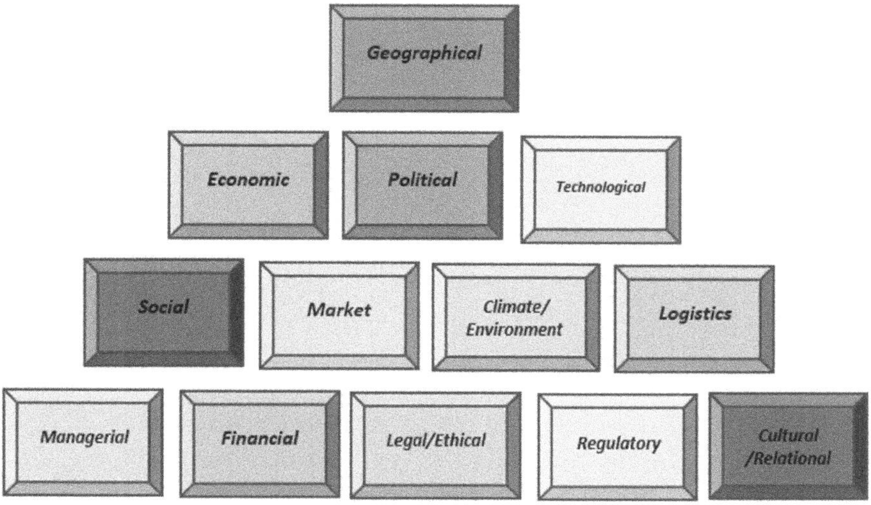

Fig. 1 Sources of risks related to supply chains (Seminar 2010)

Based on the extent of safety risks and vulnerabilities determined by the preceding steps, suitable mitigation measures are then developed and scheduled. In particular, a timetable ought to be established for the development of a plan of action with responsibilities for specific sub-plans assigned to individuals, and implementation procedures and expected outcomes documented. For instance, the use of a Radio Frequency Identification (RFID) system helps to detect potential security and safety breaches. This system can automatically collect information associated with product movement (e.g., expected time of arrival, port of entry/exit), and port authorities or cargo controllers can match received facts with pre-recorded/information documents (e.g., shipping documents, manifests). This enables them to identify suspicious trends (e.g., departures from high-risk nations), and red-flag incoming items (Min 2012). As a result, gateway security may be achieved through the early detection of anomalies that regularly result in safety or security failures.

4. *Setting up plans to govern and monitor risk reduction efforts* Each plan needs to be evaluated for its effectiveness and its impact on safety and security. In addition, it is essential to monitor the implementation of the plan and determine whether the scheduled milestones have been reached and whether progress is being made. This step is designed to improve the efficiency of security measures. By integrating blockchain technology with the mentioned steps, we can enhance security, strengthen connectivity, and ensure the safety of the supply chain. This powerful combination acts as a shield against security threats, providing an impregnable fortress to protect the integrity of the supply chain (Min 2019). Blockchain is hack-proof, tamper-proof, and irreversible because of its distributed ledger and network verification system (Karame 2018; Xu 2018).

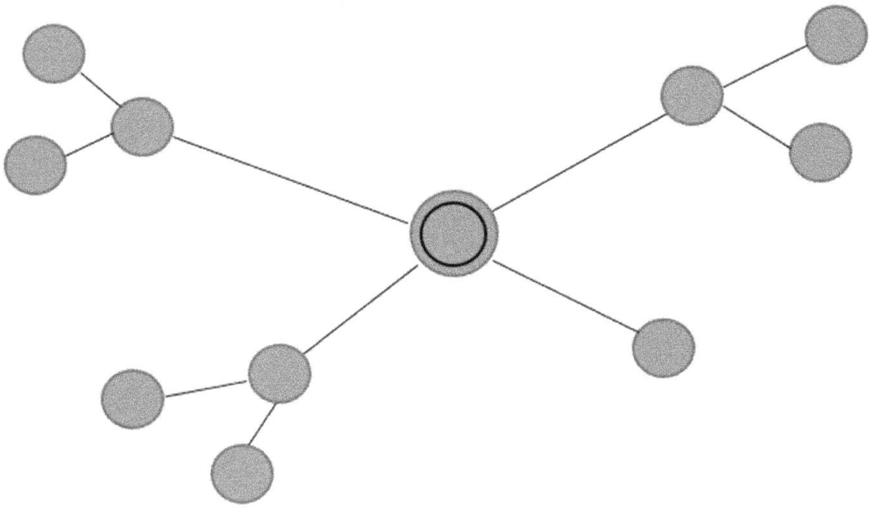

Fig. 2 The structure of supply chains with two kinds of potential risks (Grewal 2021)

Blockchain also provides automatic detectability, as an attach-only database of transaction data can be stored through a P2P network, and those historical facts will always remain by means of a fingerprint. In addition to this, a blockchain comprising nodes and arcs can be included within the ordinary supply chain structure (Fig. 2), and may be applied to capture each organizational and network risk related to the supply chain. A blockchain-based framework may be used to identify, analyze, and evaluate supply chain disruption elements and drivers based on empirical analysis of the four major disruption factor categories including (a) natural, (b) human-made, (c) system injuries, and (d) financials with drivers of disruption diagnosed and tested in a real-world industrial setting (Paul P. 2021) Significantly, a growing number of researchers and practitioners have placed supply chain resilience (SCRE) at the top of their agendas due to the increased susceptibility of global supply chains to disruptive events (Chowdhury 2017).

In order for blockchain to handle potential threats to a supply chain and improve its resilience, it is vital to know the unique activities of the supply chain network. The resilience of a supply chain may be strengthened by taking the following measures:

1. Reducing or eliminating the potential risk by identifying its source. For example, routes that are prone to maritime thefts, and ports with a high rate of theft or strikes should be avoided.

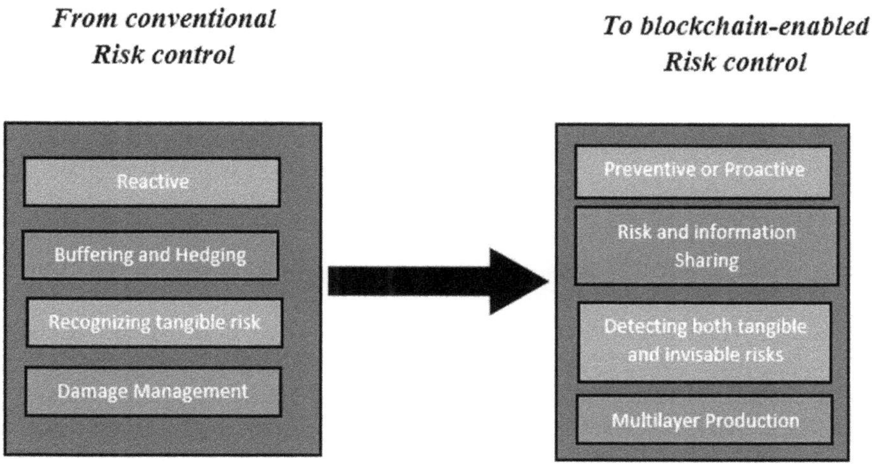

Fig. 3 Transition from conventional risk control to Blockchain-enabled risk control (Min 2012)

2. Reducing the effects of supply chain infractions by buffering with extra safety inventory, avoiding fuel fee rises for transport vehicles, and ensuring adequate insurance coverage.
3. Improving the resilience of supply chains to ensure quick recovery from infractions or unexpected crises or disasters. For instance, moving supply assets such as transportation vehicles closer to production factories.
4. Eliminating unacceptable behaviors resulting from the organization's mindset such as lax risk control practices and assumptions that predictions and forecasts are dependable, and accidents uncommon. Supply chain experts must remain open to new ideas like blockchain technology, as organizational resistance can stifle modern supply chain risk management.
5. Resilience is an essential enabler of Supply Chain Performance (SCP) because a resilient supply chain prevents disruptive impacts and assists in setting up and preserving appropriate levels of overall performance (Chowdhury 2019).

Many organizations that depend on traditional risk control methods are unaware of hidden risks such as cyberattacks, system hacking, forgery, credit failures, and contract fraud. Traditional forms of risk control have been applied to reduce obvious risks such as terrorism, theft, and natural disasters. However, these methods tend to be passive rather than proactive, and therefore, address risk management only after the event. By scrutinizing the P2P network, blockchain helps to identify and reduce risks that cannot be detected easily by supply chain members (e.g., sellers, buyers, financial institutions) during normal business transactions or supply chain activity. During various phases of a transaction, blockchain permits users to use multi-layered security scales. Figure 3 shows the traditional risk-control strategies and those enabled by blockchain.

4.2 Smart Contracts for Enhancing Supply Chain Resilience

In Sects. 4.2.1–4.2.4, we discuss how smart contracts based on blockchain technology may be applied to various sections of the supply chain to improve supply chain resilience.

4.2.1 Smart Contracts

The establishment of a transaction agreement is one of the first activities that occur in a supply chain, and contractual disputes can arise from fraudulent behavior and misunderstanding. Moreover, overall performance failures not only have the potential to ruin business partnerships, but they can also disrupt all activities within a supply chain if timely action is not taken to avoid far-reaching consequences. One of the most practical solutions to these supply chain issues is the smart contract. The smart contract is a computer-managed protocol intended to facilitate, confirm, or enforce contractual obligations by embedding contractual clauses (e.g., collateral, bonding, delineation of assets rights) within the system and then automatically executing the agreement (Szabo 1997). Hence, smart contracts not only establish the guidelines and conditions for a contractual agreement, but they also immediately establish the responsibilities of all parties involved. They are self-checking and self-running agreements that automate the settlement lifecycle to mitigate risk, and improve the effectiveness and efficiency of a company's transactional processes (Icertis 2017). Smart contracts can be converted to system codes that are saved and copied on the computer system and monitored with the aid of the blockchain that has been integrated into the system. Smart contracts can facilitate transparent transactions involving money, property, stocks, or any valuable item or product without the need for the services of intermediaries (Blockgeeks 2017). Hence, smart contracts are autonomous, thereby reducing transaction times and costs. Moreover, by integrating the Internet of Things (IoT) into the blockchain, contract fraud can be discovered easily, and/or avoided. Smart contracts make blockchain applications more operationally comfortable and economically attractive. Giovanni (2020) Additionally, as shown in Fig. 4, the integrity of asset transfers established by the agreement is improved because the decentralized network has many members and there is more robust security.

4.2.2 Asset Tracking

Smart contracts can be used to track asset ownership. As soon as both tangible and intangible assets are recorded on the blockchain, ownership is permanent until the proprietor confirms otherwise. The paperless record created by the blockchain is immutable. In addition, blockchain works as a perfect and easily distinguishable record that completely tracks and saves all the supply chain-associated obligations

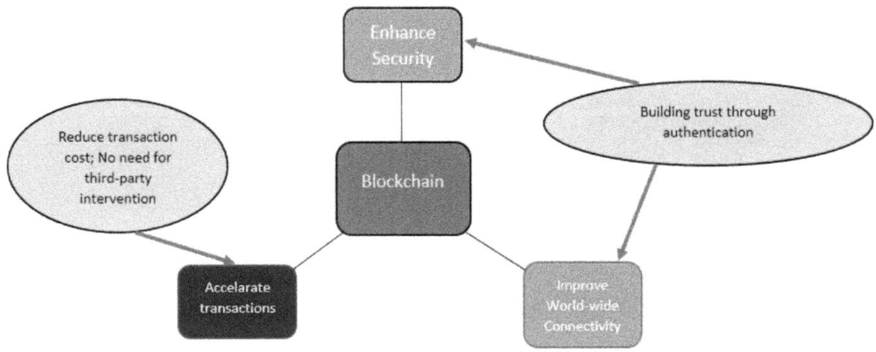

Fig. 4 Combination of blockchain and IoT (Giovanni 2020)

for a particular resource. Blockchain permits transparency all the way back to the origin of an asset. Therefore, blockchain prevents the exchange of false properties and makes it much easier to track objects as they are made and conveyed along the supply chain. In the U.S., the Treasury office follows and oversees the movement of physical resources in real time as they move from one party to another all along the supply chain, utilizing virtual information of exchanges of possessions recorded on blockchain (Higgins 2017). Furthermore, blockchain may be utilized to receive shipments in global logistics operations. For example, Maersk, the Danish integrated logistics company recently completed a 20-week blockchain proof-of-concept to track cargo movements. Blockchain technology's cryptographic signatures serve as a powerful safeguard, preventing unauthorized changes to shipping details like names and cargo identification. Furthermore, this integration minimizes the chances of lost shipments during transit, creating a streamlined and trustworthy supply chain environment that operates with greater efficiency (Min 2019; Green 2017; CBS 2015). Companies can also enforce a robust logistics system to control last-mile delivery and secure better overall performance through blockchain (Naclerio 2022).

4.2.3 Safe and Accurate Order Delivery

With easy access and no need for paper-based client information, blockchain will expedite order fulfillment through the supply chain by means of high-speed confirmation of a purchaser's credit history, stock repute, budget, and report on order/delivery popularity while delivering visibility during the order fulfillment process. As shown by the circles in Fig. 5, through the mechanization of the rotated order fulfillment stages, blockchain might reduce back-order fulfillment errors and increase the pace of the order fulfillment procedure. Moreover, because the blockchain ledgers are transparent and can be visited by any peer-to-peer network player (e.g., each customer and vendor), blockchain visibility will improve the transparency of the order fulfillment process and mitigate the risk of fulfillment errors.

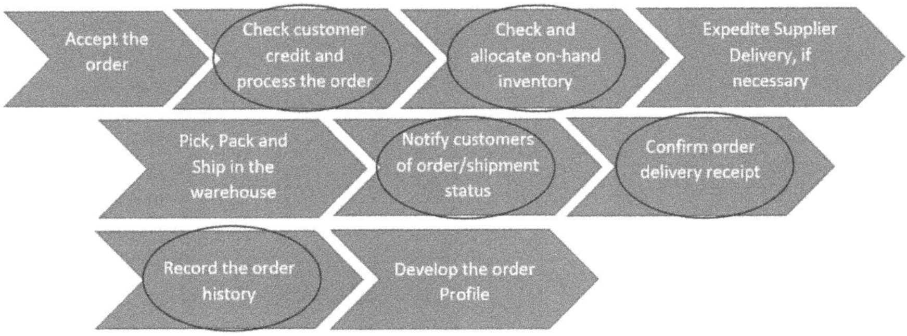

Fig. 5 Part of order fulfillment stages where blockchain technology can be applied (CBS 2015)

4.2.4 Cybersecurity

The last decade has seen a significant increase in cybercrime such as the hacking of the international meat processor JBS and the US oil distribution giant Colonial Pipeline (Naclerio 2022; Oloruntoba and Thompson 2021; Statista 2016). Cybercrime can cripple supply chain processes at any point in the supply chain network. However, despite cybersecurity measures such as antivirus, password security, and hazard cautions to address this issue, the risk of cybercrime has in no way lessened. Efforts at fighting cybercrime may utilize blockchain with its end-to-end encryption, immutability, and privacy, and could reduce the risk of system failure. The immutability of blockchain and the fact that each computer in a P2P network constantly checks data before storing it on the blockchain makes it a high-quality tool to reduce the risk of cybercrime and hacking. Thus, blockchain guarantees the integrity of the transaction record, and nobody, including the record proprietor, can alter the account as soon as the account is locked into the blockchain.

5 Blockchain Technology Applications for Resilience

Global supply chains require a technique that could help to organize facts reliably and provide tools for risk evaluation and disruption mitigation. Such risk management strategies can be constructed with the use of blockchain. Firstly, created to facilitate Bitcoin cryptocurrency exchanges, blockchain consists of a distributed ledger where transactions are patched to more than one computer rather than being stored in one central location. The transparency of transactions involving several parties provides straightforwardness and security. Once records are transferred to the blockchain, they cannot be altered, thereby ensuring permanence and trustworthiness. This becomes the keystone of a sound, computerized supply chain. When transactions are stored in blockchain, all relevant data are obvious to all supply chain entities. The permanence, traceability, and straightforwardness reveal the provenance of items along the entire supply chain.

5.1 *Visibility*

With the advent of the COVID-19 pandemic, the transparency of most global supply chains began to be questioned in 2020. During the pandemic, a company's sustainability depended on how well it knows and can see all aspects of its value chain. Supply chain structures can collapse under the significant stress of catastrophic global events. Many companies were unaware of the amount of upstream and downstream stock they had, which would enable them to modify demand in a timely fashion. If one company in the supply chain fails, this presents a risk to all other related companies. However, they did not understand enough about the network to realize that the new first-tier provider relies upon the identical old second-tier provider. Moreover, many of those corporations are located within the Chinese epicenter of the current pandemic. These "invisible" producers have had to quickly shut down, or have become understaffed because of quarantine regulations (Tom Linton 2020), which in turn greatly affects companies and supply chains across the world. A supply chain based on blockchain connects all parties within a network, unlocking complex worldwide structures and providing transparency at each step and in each link of the supply chain. Companies can view all tiers, locations, and production facilities of suppliers and subcontractors. Blockchain allows them to evaluate risk, simulate scenarios, perform what if evaluations, and take preventive actions. Additionally, businesses can respond rapidly to unexpectedly changing conditions. Real-time records on the precise location of a product offer businesses more flexibility and control. As an example, a Swiss company orders 100,000 devices, half of which have been made in Japan and half made in Wuhan, China. On the blockchain, the company notices a slowdown in manufacturing in Wuhan. This permits the organization to take the correct action to minimize any adverse impacts.

5.2 *Digitization*

The global shipment of goods decreased by an estimated 70% during the pandemic, and 40% of China's freight traffic was shut down because of quarantine regulations and fears of spreading infection. Countries have applied many precautions to contain the virus, including banning ships' crew members from disembarking (Ship-technology.com 2020). A few ports required ships to declare that all crew members were healthy and did not have the virus. If COVID-19 was suspected on board, the ships could not dock in port (portofantwerp.com 2020). Despite these precautions, port employees and truckers were still at great risk of infection due to their interactions with people from other regions or countries. A basic weakness of many global commercial enterprises is the lack of end-to-end digitization of various files and procedures within the supply chain, which makes it impossible to function without physical interaction. The level of human interplay in a supply chain creates the trust engendered by the use of specific documentation and/or techniques. Individuals, organizations, nations, and the supply chain need to improve the virtual and digital capabilities of their existing supply chains by establishing trustworthiness, which is not always a feature of business relationships as physical contacts and traditional methods go out of date. Trustworthiness ought to be system-specific. It needs to be scalable to support the supply chain in all circumstances. In this regard, blockchain technology can help enterprises to move toward a trusted virtual supply chain. The blockchain is a reliable virtual database that stores records of transactions between all entities. In this manner, the blockchain offers all participants a single source of information. The shipment of a single ocean container typically involves the participation of approximately 28 different parties, each having their own distinct data storage structures. Even before a ship arrives at the port, the government receives all information pertaining to the owner, shipment contents, team, and direction, all of which require a substantial amount of paperwork, which can be eliminated by using blockchain technology. Port authorities can then determine whether the delivery needs to be checked, quarantined, or cleared. This diminishes uncertainty and the need for human intervention. Blockchain has many advantages and has been implemented in many domains worldwide. Rotterdam, one of the busiest ports in Europe developed a blockchain-based solution for issues related to port logistics. The whole delivery process is document free (portofantwerp.com 2019). In the US, deliveries and orders are transparent to all parties in real time, and monetary transactions are concluded immediately, reducing danger and increasing reliability and interoperability using the US government blockchain in customs control. In the US, due to COVID-19, many logistics personnel have been working from home. This has been facilitated by blockchain which can handle transactions and data storage securely, so customs authorities are not physically present in ports (Gillis 2020). Smart contracts automate digital transactions and establish agreements and conditions. The reconciliation of cargo information can be a labor-intensive and time-consuming procedure, particularly in times of adversity. This places an economic burden on shipping businesses which are already prone to losing their business because of port and government regulations and occasional high volume of demand. Smart contracts may be beneficial in those situations. Through a smart contract, the bill of lading is digitally signed and established as soon as the truck transports the products to the warehouse, and payment is immediately transferred to the shipping company.

Transactions are immediately visible to all stakeholders such as banks, retailers, and shipping corporations. Hence, blockchain technology is of great value in this regard.

5.3 Provenance

Over the last decade, businesses have been pressured to provide dependable and visible supply chain processes, and never more so than during the COVID-19 pandemic. Clients worry that a product might have come from or passed through a place where the virus is rampant. Customers, practitioners, and governments now confirm the location of manufacturers and where they are within the supply chain, and whether or not goods might have been contaminated, spoiled or infected. As an example, consider meat being shipped in a refrigerated container. As soon as the meat arrives in the warehouse, the company takes numerous samples, and the quality is checked in a laboratory. Even though the products may be of high quality, the host organization might not be confident about the origin of the meat or the delivery process of the entire order. The only guarantee of the quality of the meat is the word of the logistics company that the refrigerated storage chain was as promised throughout the delivery process. These concerns can be addressed with the help of blockchain which gives transparency to the activities of various parties such as producers, shipping businesses, cargo transport companies, wholesalers, and retailers who are members of the same blockchain system. When counting the meat items or assessing their quality, the company stores this data and uploads it to the blockchain community. Next, the meat is packaged, transferred to refrigerated packing containers, and transported via sea to diverse areas. Upon arrival at the port, the meat products are transported by refrigerated vehicles to the wholesalers' warehouses from whence it is distributed to retailers. During the journey, tamper-resistant IoT devices within the storage holders and trucks can record temperatures and send readings to the blockchain on an hourly basis. Smart contracts may specify that when the temperature exceeds or falls below a certain level, an immediate notification is sent to relevant parties to warn them of these variations. Decisions can then be made about potentially discarding the product. When the product is bought, purchasers can track the entire product journey by simply scanning the QR code. Knowing the provenance of a product is particularly important in the case of food and other perishable products and food items in order to avoid poisoning. Products from affected areas can be quickly disposed of, and merchandise from secure regions can be shipped as usual. For example, in mid-March 2020, as the COVID-19 pandemic struck the USA with unintended effects and scale, hospitals struggled to ensure an adequate supply of crucial private protective equipment (PPE) that they needed on the front lines. So it is essential to study how corporations in the global food value chains managed their supply chain structures in reaction to the COVID-19, turning it more resilient and competitive in the present pandemic and in future disruptions (Ali 2022; Sanjoy and Moktadir 2021).

6 Role of Blockchain Technology in Mitigating Pandemic-Induced Supply Chain Disruptions

There are five main ways that blockchain helped to address supply chain disruption during the pandemic. These are discussed below.

1. Trust is fundamental because the pandemic hit the supply chain, inflicting shortages, rising costs, and great disruptions to factories, ports, suppliers, and purchasers, customers want a system they can trust. Blockchain is well-known for being a trustworthy system that can operate independently of a person or agency. The trustworthiness of blockchain lies in the fact that it creates an environment consisting of third-party validators, immutable data, unchangeable logic, and supplier information and traceability.
2. Blockchain establishes a foundation of trustworthiness; however, the natural environment brings it to life. Every blockchain establishes a new industry environment in which individuals collaborate and cooperate. But, when trustworthy industry partners participate as validators and solution providers within the industry sector, the advantages are numerous and give the industry additional vitality.
3. Inventory transparency ensures compliance with regulations and the timely transport and delivery of goods. During a pandemic, incorrect information about the most important items that should be delivered can result in severe losses, including loss of lives when there are shortages of essential items. Blockchain can offer real- time transparency of crucial supply chain processes such as procurement, production, transportation, and distribution.
4. The logistics and supply chain domain need a simple and quick solution to problems caused by disruptions such as the pandemic. Supply chain solutions may be very complicated, as massive consortia consisting of numerous members require a single source of trust. Solutions to supply chain problems cannot be created in a single day when an epidemic strikes. To address marketplace issues quickly, market-established and available solutions that have been applied for years are utilized. However, the extent of consolidation may be intimidating and can take a long time. Blockchain can simplify solutions and address supply chain issues quickly, or even prevent them.
5. Trusted digital identification bridges the trustworthiness gap between the digital world and the actual world. Blockchain-based digital identities play a critical role in building trust. Digital identification is defined as the connection between one's real identification and the digital world. A digital copy of any physical item (called a digital twin) is produced in a way that hampers or detects human attempts to change records. For digital business transactions within the supply chain, it is crucial to verify the authenticity of an identity and confirm the identity of one's transaction partner. During the pandemic, other functions of blockchain can help to strengthen supply chain resiliency. These features include ensuring the privacy of shared records, progressive industrial venture styles that bring partners together within the aggressive natural environment, and third-party

arrangement integrations that work on top of those ecosystems. With numerous supply chains now adopting blockchain, and interoperability being addressed effectively, society can become more resilient against the consequences of the current pandemic and future disruptions.

6.1 Discussion

The COVID-19 pandemic revealed the fragility of global supply chains. Companies need to thoroughly identify and assess problems throughout the recovery process and formulate strategies that will ensure the survival of their organizations and supply chains (Sanjoy 2021). Blockchain technology offers immutable record-keeping of all data (Adeodato 2020). Blockchain technology acts as a safeguard by archiving all shared data within a network, effectively preventing interruptions and data breaches commonly encountered through network attacks. Through decentralization, the risk of a single point of failure is dispersed, while the traceability feature of permanent record-keeping, supported by digital signature functions, eliminates the risk of insider attacks originating from individual vector (Bayramova 2021). This is because only authentic customers are permitted in the network. However, in spite of these capability benefits such as the strengthening of resilience against more than one disruption in the 'physical' and 'digital world', there are various challenges related to blockchain implementation for each vendor and each adopter. This requires developing guidelines (as a theoretical framework) that can be used by potential blockchain adopters and vendors. To avoid discrepancies between organizational needs and technological offerings, it is crucial for potential adopters of blockchain technology to clearly define their requirements. By doing so, they can effectively identify and choose appropriate blockchain applications and functionalities that align with their needs, as emphasized by Epiphaniou (2020). Furthermore, vendors are responsible for assessing the compatibility of companies' requirements with the available blockchain solutions, as highlighted by Kshetri (2019). In order to gauge the return on investment (ROI), it is essential to consider the potential value and costs associated with the latest technological solutions, as emphasized by (Lu 2017; Bayramova 2021). For example, adopting a traceability system is profitable for high-value goods (such as diamonds) as a means of preventing the counterfeiting of products (Kshetri 2019) or the selling of products that are perishable, such as medicines with use-by dates, or dairy and meat products that have been spoiled (Lu 2017). Hence, software program developers or vendors should verify the feasibility and degree of adaptability of the adopter's current system to ensure the successful integration of the proposed blockchain solution (Erol 2021). Ultimately, adopters must take into account the interoperability of their selected blockchain solution within supply chains, as the merits of this disruptive technology are leveraged through broad-based acceptance instead of being limited to only one supply chain enterprise (Sternberg n.d.). Furthermore, vendors must examine the degree of technological maturity of potential adopters in order for the blockchain technology to be successful (Erol 2021; Sternberg n.d.).

6.2 Managerial Benefits and Challenges of the Blockchain Technology

In addition to providing greater security as discussed in the previous sections, blockchain can offer a large number of managerial advantages to regular business practices (Techlab 2017; Takahashi 2017), which include: 1. Reduced transaction fees/time as a result of better-preserved blockchain structures that do not necessitate third-party involvement. 2. Improvement of visibility throughout the supply chain, resulting from the increased transparency of open ledgers accessible to all members. 3. The merging of digital and physical realms in the supply chain, as highlighted by Techlab (2017) and Takahashi (2017), leads to enhanced connectivity among trading partners. This integration enables the sharing of transactional information and data across the supply chain, promoting transparency. By utilizing advanced technologies like electronic data interchange (EDI), extensible markup language (XML), and application programming interface (API), blockchain paves the way for swift and traceable interactions, enabling the exchange of unchangeable records among supply chain collaborators, as emphasized by IBM (2017) and Min (2019).

Despite the many advantages of blockchain discussed in the previous section, it has some inadequacies and execution issues mainly due to its complexity, newness and revolutionary technology. A number of those issues are depicted in Fig. 6, the most pressing of which is associated with adaptability, interoperability, and administration.

Each node must process and verify each transaction in a blockchain, so blockchains inherently require massive computing strength and high-speed internet connections, which are not easy to construct even with state-of-the-art technology. If the blockchain centralizes the verification method, such an approach could defeat its unique purpose. In addition, the different platforms available for blockchain make it challenging to locate the optimal mixture of various structures and platforms that are interoperable. Because blockchain relies on a distributed ledger that could evade authorities' intervention, governments can put pressure on blockchain customers by imposing regulatory and criminal limitations. Therefore, this might impede blockchain's ability to guarantee the integrity and privacy of transactions and the movement of assets (De 2018). Unexpectedly, the delivered privacy, confidentiality, and security can make it extremely difficult for law enforcement authorities to choose who oversees digital wallets, making them more vulnerable to scammers looking to steal computerized monetary transactions recorded on the blockchain (Hackett 2017).

7 Summary, Conclusions, and Future Research Directions

Discussions in this chapter suggest that blockchain technology is still in the embryonic stage, although it is likely to evolve to meet societal, economic, and political needs and applications. However, currently, blockchain technology has yet to be widely implemented in supply chains in real-world contexts. Most case examples

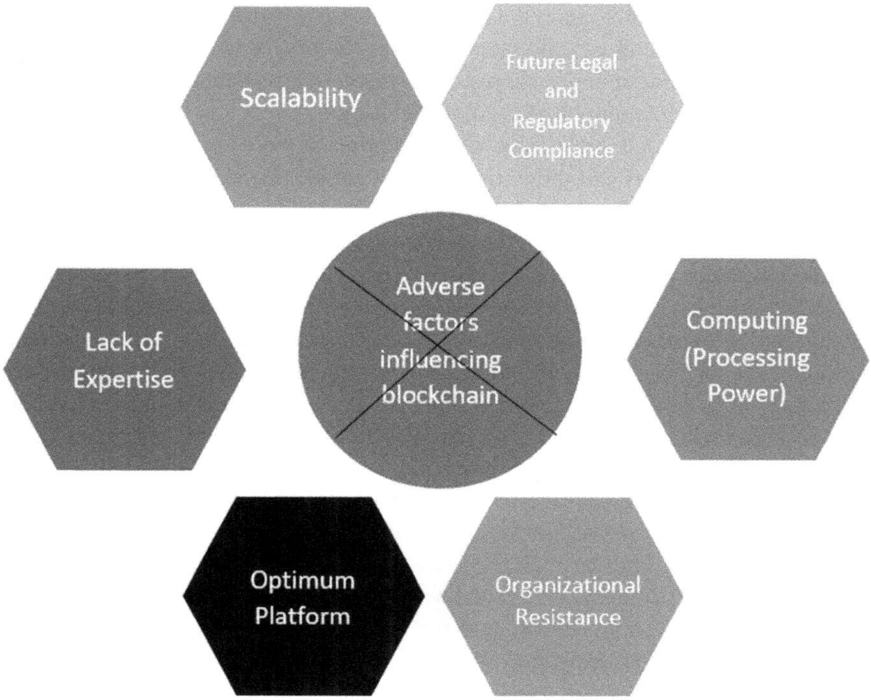

Fig. 6 Potential challenges of blockchain technology (De 2018)

lack standard strategies for designing and validating the blockchain solutions (Perboli 2018). As indicated by Deloitte Insights (2017), a mere 8% of blockchain projects have endured, leaving the majority as unsuccessful endeavors. This lack of success can potentially be attributed to individual users developing isolated blockchain applications, rather than focusing on creating foundational libraries that enable the development of multiple applications (Insights 2017). Furthermore, corporate-led projects tend to incur higher adoption costs compared to user-driven initiatives, as organizational blockchain projects are reportedly five times more likely to be replicated by competitors (Insights 2017; Bayramova 2021). The chapter has discussed the latest trends and filled gaps in the knowledge of blockchain technology. The chapter will enable both researchers and practitioners to better understand the blockchain technology landscape and trends. Further contributions of the chapter include highlighting the practical implications of blockchain for cyber risk management, and the need for guidelines that specify the necessities for blockchain software developers, vendors, and providers. The chapter also highlights the value of blockchain technology as a means of improving trust and time and cost- effectiveness within the supply chain, and strengthening the overall supply chain's resilience during disruptions, such as those caused by the COVID-19 pandemic. The chapter examined the role of smart contracts in enhancing supply chain resilience. It then examined the role of blockchain technology in the building blocks of resilience such as visibility, digitization, and provenance, and then examines the role of blockchain technology in mitigating pandemic-induced supply chain risks, and the benefits and managerial challenges of implementing blockchain technology.

Because blockchain is a relatively new development, there has been little research on it, and the concept is still unfamiliar to most people. Hence, the purpose of this chapter is to establish the conceptual framework for blockchain from a business perspective. The chapter discusses the increasing implementation of blockchain and identified the value of blockchain for supply chain management from a risk/security perspective using commonly accepted supply chain practices and processes. This chapter can be expanded to include and illustrate successful case studies of real-world blockchain programs while evaluating the effects of situational variables such as company size, functional department, budget sizes, and organizational readiness for blockchain adoption (Ashraf 2014; Cooper 1990; Davis 1989; Maruping 2017; Mathieson 1991; Oliveira 2011). Future studies may need to expand the scope of research to determine the causal relationships between these variables and the selection of the most appropriate blockchain solution, while taking into consideration cross-cultural and/or long-term factors. Also, future studies need to investigate the potential effects of different contextual variables along with industry traits, peer pressure, government intervention, and the degree of senior management support for blockchain adoption decisions. Other studies could compare the alignment of blockchain with supply chain resilience that incorporates business insights, cloud computing, robotics, and/or artificial intelligence (AI) (Min 2012). Technologically, blockchain can be an excellent solution to supply chain issues as it provides essential transparency, perceivability, tracking, and trust. In fact, it is emerging as a modern approach to building trust in trust-free surroundings while also providing assurance of availability and security in data management. Dwivedi (2022) It is simple to integrate blockchain with existing enterprise asset-planning systems and supply chain structures. Blockchain levels the playing field by creating an open market and taking control from people who in the past have taken advantage of the weaknesses of current systems. However, on the downside, the implementation of blockchain can cause major disruption to current processes and it could meet with resistance from some supply chain actors.

Some may resent transparency, fearing that blockchain could reveal their network's vulnerabilities and risk factors to their clients. Also, acceptance of, and adaptation to, change can be difficult. However, if supply chain stakeholders or their representatives utilize this opportunity to develop more effective systems, the global supply network can become more robust and much better suited to the volatile requirements of our modern-day world. Customers can be the driving force of this change. Implementation can be facilitated by establishing a consortium of stakeholders in the supply chain network for the purpose of discussing the benefits and demerits of implementing blockchain in global networks. This is because it is likely that there will be other unexpected and unavoidable disruptions, making it essential to take ambitious steps now to design and implement an ultra-resilient supply chain model for the future.

References

Aceto B (2019) Blockchain e Dintorini. http://tendenzeonline.info/articoli/2019/05/08/blockchain-edintorini

Adeodato R, Pournouri S (2020) Secure implementation of e-governance: a case study about estonia. Adv Sci Technol Secur Appl

Ali I (2022) Reimagining global food value chains through effective resilience to COVID-19 shocks and similar future events: a dynamic capability perspective. J Bus Res 141:1–12

Angrish CBHMSB, A (2018) A case study for blockchain in manufacturing, FabRec: a prototype for peer-to-peer network of manufacturing nodes. Procedia Manuf 26:1180–1192

Armstrong S (2018) Bitcoin technology could take a bite out of NHS data problem. BMJ: Br Med J 361

Ashraf AR, Thongpapanl N, Auh S (2014) The application of the technology acceptance model under different cultural contexts: the case of online shopping adoption. J Int Mark 22(3):68–93

Bailey G (2020) Coronavirus and the remaking of global supply chains. Forbes.com

Bayramova A, Edwards DJ, Roberts C (2021) The role of blockchain technology in augmenting supply chain resilience to cybercrime. Buildings 11(7):283

Blockgeeks (2017) Smart contracts: the blockchain technology that will replace lawyers. https://blockgeeks.com/guides/smart-contracts/

Boyes H (2015) Cybersecurity and cyber-resilient supply chains. Technol Innov Manag Rev

Butt AS (2021a) Supply chains and COVID-19: impacts, countermeasures and post-COVID-19 era. Int J Logistics Manag

Butt AS (2021b) Understanding the implications of pandemic outbreaks on supply chains: an exploratory study of the effects caused by the COVID-19 across four South Asian countries and steps taken by firms to address the disruptions. Int J Phy Distribu Logistics Manag

Casey M, Wong P (2017) Global supply chains are about to get better, thanks to blockchain. Harv Bus Rev 13

CBS. (2015). These cybercrime statistics will make you think twice about your password: where's the CSI cyber team when you need them?. http://www.cbs.com/shows/csi-cyber/news/1003888/these-cybercrimestatistics-will-make-you-think-twice-about-yourpassword-where-s-the-csi-cyber-team-when-youneed-them-/

Chakrabarti A, Chaudhuri A (2017) Blockchain and its scope in retail. Int Res J Eng Technol 4

Chen A-N, T.M (2011) Lessons from stuxnet. Computer

Chowdhury MM, Quaddus M (2017) Supply chain resilience: conceptualization and scale development using dynamic capability theory. Int J Prod Econ

Chowdhury Q-M, MMH (2019) Supply chain resilience for performance: role of relational practices and network complexities. Int J Supply Chain Manag

Christopher P-H, M (2004) Building the resilient supply chain 2(2)

Christopher M (2000) The agile supply chain: competing in volatile markets. Ind Mark Manag 29(1):37–44

Cooper ZRW, RB (1990) Information technology implementation research: a technological diffusion approach. Manag Sci 36(2):123–139

Cox CD, A (2005) The limits of lean management thinking: multiple retailers and food and farming supply chains. Euro Manag J 23(6):648–662

Crosby M, Pattanayak P, Verma S, Kalyanaraman V (2016) Blockchain technology: beyond bitcoin. Appl Innov 2:6–10

David M (2003) How is the SARS virus infecting the supply chain? will you be affected? electronicdesign.com

Davis F (1989) Perceived usefulness, perceived ease of use, and user acceptance of information technology. MIS Q 13(3):319–339

De N (2018) Report: South Korea eyes joint crypto regulations with China, Japan. https://www.coindesk.com/report-south-korea-eyesjoint-crypto-regulations-with-china-japan/

Deloitte (2019) The future of cyber survey 2019 cyber everywhere. Succeed Anywhere; Deloitte, 1–32
Dutra A, Tumasjan A, Welpe IM (2018) Blockchain is changing how media and entertainment companies compete. MIT Sloan Manag Rev 60(1):39
Dwivedi A, Agrawal D, Paul SK, Pratap S (2022) Modeling the blockchain readiness challenges for product recovery system. Ann Oper Res
Epiphaniou G, Bottarelli M, Al-Khateeb H, Ersotelos NT, Kanyaru J, Nahar V (2020) Smart distributed ledger technologies in industry 4.0: challenges and opportunities in supply chain management. Adv Sci Technol Secur Appl
Erol I, Ar IM, Ozdemir AI, Peker I, Asgary A, Medeni IT, Medeni T (2021) Assessing the feasibility of blockchain technology in industries: evidence from Turkey. J Enterp Inf Manag 34:746–769
Filimonau V, Naumova E (2020) The blockchain technology and the scope of its application in hospitality operations. Int J Hosp Manag 87
Ghadge A, Wei M, Caldwell ND, Wilding R (2019) Managing cyber risk in supply chains: a review and research agenda. Supply Chain Manag
Gillis C (2020) More US customs broker, forwarder employees work from home. freightwaves.com
Giovanni P (2020) Blockchain and smart contracts in supply chain management: a game theoretic model. Int J Prod Econ 228
Glickman TS, White SC (2006) Security, visibility and resilience: the keys to mitigating supply chain vulnerabilities. Int J Logist Syst Manag 2(2):107–119
Green A (2017) Will blockchain accelerate trade flows?. Financ Times. https://www.ft.com/content/a36399fa-a927-11e7-ab66-21cc87a2edde
Grewal D, Gauri DK, Das G, Agarwal J, Spence MT (2021) Retailing and emergent technologies. J Bus Res 134:198–202
Guo Y, Liang C (2016) Blockchain application and outlook in the banking industry. Financ Innov 2(1):24
Hackett R (2017) Maersk and microsoft rested a blockchain for shipping insurance. Fortune. http://fortune.com/2017/09/05/maersk-blockchain-insurance/
Higgins S (2017) The US treasury is testing distributed ledger asset tracking. https://www.coindesk.com/us-treasury-testing-distributed-ledger-asset-tracking/
IBM (2017) The benefits of blockchain to supply chain networks. IBM Corporation, Somers, NY
Icertis (2017) Smart contracts are transforming the way we do business. https://www.icertis.com/resource/smart-contracts-are-transforming-the-waywe-do-business-featuring-gartner-research
Insights D (2017) The evolution of blockchain technology. https://www2.deloitte.com/us/en/insights/industry/financial-services/evolution-of-blockchain-github-platform.html
Ivanov D, Dolgui A (2021) A digital supply chain twin for managing the disruption risks and resilience in the era of industry 4.0. Prod Plan Control 32(9):775–788
Kache F, Seuring S (2017) Challenges and opportunities of digital information at the intersection of big data analytics and supply chain management. Int J Oper Prod Manag 37110–37136
Karame G, Capkun S (2018) Blockchain security and privacy. IEEE Secur Priv 160411–160412
Kshetri N (2017) Blockchain's roles in strengthening cybersecurity and protecting privacy. Telecommun Policy 41:1027–1038
Kshetri N, Loukoianova E (2019) Blockchain, adoption in supply chain networks in Asia. IT Prof 21:11–15
Langmann B, Folch A, Hensch M, Matthias V (2010) Volcanic ash over Europe during the eruption of Eyjafjallajökull on Iceland. Atmosp Environ 48:1–8
Linton T, Vakil B (2020) Coronovirus is proving we need more resilient supply chain. HBR.org
Lu Y (2019) The blockchain: State-of-the-art and research challenges. J Ind Inf Integr 15:80–90
Lu Q, Xu X (2017) Adaptable blockchain-based systems: a case study for product traceability. IEEE Softw 34:21–27
Mainelli M, Smith M. Sharing ledgers for sharing economies: an exploration of mutual distributed ledgers (aka blockchain technology)

Marr B (2018) 30 Real examples of blockchain technology in practice. www.forbes.com/sites/bernardmarr/2018/05/14/30-real-examples-of-blockchain-technology-inpractice/212e860d740d
Maruping LM, Bala H, Venkatesh V, Brown SA (2017) Going beyond intention: integrating behavioral expectation into the unified theory of acceptance and use of technology. J Assoc Inf Sci Technol 68(3):623–637
Mathieson K (1991) Predicting user intentions: comparing the technology acceptance model with the theory of planned behaviour. Inf Syst Res 2(3):173–191
Meier M, Pinto E (2020) COVID-19 Supply chain disruptions CRC TR 224 discussion paper series. University of Bonn and University of Mannheim, Germany
Min H (2019) Blockchain technology for enhancing supply chain resilience. Bus Horizo 62(1):35–45
Min H, Shin SS (2012) The use of radio frequency identification technology for managing the global supply chain: an exploratory study of the Korean logistics industry. Int J Logist Syst Manag 13(3):269–286
Misra P (2018) 5 ways blockchain technology will change the way we do business. www.entrepreneur.com/article/309164
Naclerio AG, De Giovanni P (2022) Blockchain, logistics and omnichannel for last mile and performance. Int J Logist Manag 33(2)
Nandi S, Sarkis J, Hervani AA, Helms MM (2021) Redesigning supply chains using blockchain-enabled circular economy and COVID-19 experiences. Sustain Prod Consum 41:10–22
Oliveira T, Martins MF (2011) Literature review of information technology adoption models at firm level. Electron J Inf Syst Eval 14(1):110–121
Oloruntoba R, Thompson N (2021) Cyber attacks can shut down critical infrastructure. It's time to make cyber security compulsory. https://theconversation.com/cyber-attacks-can-shut-down-critical-infrastructure-its-time-to-make-cyber-security-compulsory-160991
Osgood R (2016) The future of democracy: blockchain voting. In: COMP116: information security
Pattison I (2017) 4 Characteristics that set blockchain apart. www.ibm.com/blogs/cloud-computing/2017/04/11/characteristics-blockchain/
Paul Sanjoy K (2021) Supply chain recovery challenges in the wake of COVID-19 pandemic. J Bus Res 136:316–329
Paul P, A. A. F- F. C. J. C. J. S. L, Chowdhury R (2021) Modelling of supply chain disruption analytics using an integrated approach: an emerging economy example. Expert Syst Appl 173
Peck H (2005) Drivers of supply chain vulnerability: an integrated framework. Int J Phys Distrib Logist Manag
Peck H (2006) Reconciling supply chain vulnerability, risk and supply chain management. Int J Logist: Res Appl 9(2):127–142
Perboli G, Musso S, Rosano M (2018) Blockchain in logistics and supply chain: a lean approach for designing real-world use cases. IEEE Access 6:62018–62028
Pettit TJ, Fiksel J, Croxton KL (2010) Ensuring supply chain resilience: development of a conceptual framework. J Bus Logist 31(1):1–21
Pilkington M (2016) Blockchain technology: principles and applications. In: Research handbook on digital transformations. http://ssrn.com/abstract=2662660
portofantwerp.com (2019) How Rotterdam is using blockchain to reinvent global trade. portofantwerp.com
portofantwerp.com (2020) Coronavirus-Update 9 March. portofantwerp.com
Raab M, Griffin-Cryan B (2011) Digital transformation of supply chains: creating value-when digital meets physical. Capgemini Consulting, Paris
Raval S (2016) Decentralized applications: harnessing bitcoin's blockchain technology. O'Reilly Media, Sebastopol, CA
Rejeb A, Karim R (2019) Blockchain technology in tourism: applications and possibilities. World Sci News 137:119–144

Reyna A, Martin C, Chen J, Soler E, Diaz M (2018) On blockchain and its integration with IoT. Challenges and opportunities. Fut Gener Comput Syst 88:173–190
Rivera L, Wan HD, Chen FF, Lee WM (2007) Beyond partnerships: the power of lean supply chains. Trends Supply Chain Des Manag 241–268
Sanjoy Paul KSTCRC, Moktadir Abdul (2021) A recovery planning model for online business operations under the COVID-19 outbreak. Int J Prod Res 141:1–23
Sawik T (2022a) A linear model for optimal cybersecurity investment in industry 4.0 supply chains. Int J Prod Res 60(4):1368–1385
Sawik T (2022b) Balancing cybersecurity in a supply chain under direct and indirect cyber risks. Int J Prod Res 60(2):766–782
Seminar C-T, T (2010) C-TPAT training seminar. In: C-TPAT 5 step risk assessment process guide, Washington
Sherman E (2020) 94% of the Fortune 1000 are seeing coronavirus supply chain disruptions: report. Fortune.com
Ship-technology.com (2020) Coronavirus outbreak: measures and preventive actions by ports. Ship-technology.com
Spieske A, Birkel H (2021) Improving supply chain resilience through industry 4.0: a systematic literature review under the impressions of the COVID-19 pandemic. Comput Ind Eng 158
StatistaStatista (2016) Statistics and market data on cybercrime. https://www.statista.com/markets/424/topic/1065/cyber-crime/
Sternberg HS, Hofmann E, Roeck D (n.d.) The struggle is real: insights from a supply chain blockchain case
Strozzi F, Colicchia C, Creazza A, Noe C (2017) Literature review on the "Smart Factory" concept using bibliometric tools. Int J Prod Res 55:6572–6591
Swan M (2015a) Blockchain: blueprint for a new economy. O'Reilly Med 55:6572–6591
Swan M (2015b) Climate change 2013-the physical science basis. Blueprint for a New Economy, O'Reilly Media, In Blockchain
Szabo N (1997) Formalizing and securing relationships on public networks. http://ojphi.org/ojs/index.php/fm/article/view/548/469
Takahashi R (2017) How can creative industries benefit from blockchain? McKinsey Company. https://www.mckinsey.com/industries/media-and-entertainment/our-insights/how-can-creative-industries-benefit-from-blockchain
Tapscott D, Tapscott A (2016) Blockchain revolution: how the technology behind bitcoin is changing money, business, and the world. Penguin, New York, NY
Taylor PJ, Dargahi T, Dehghantanha A, Parizi RM, Choo KK (2019) A systematic literature review of blockchain cyber security. Digit Commun Netw
Taylor D (1999) Parallel incremental transformation strategy: an approach to the development of lean supply chains. Int J Logist: Res Appl 2(3):305–323
Techlab M (2017) What is blockchain and understand its key benefits, 42. https://www.marutitech.com/blockchain-benefits/
Theguardian.com (2020) Nissan and BMW car production hit by volcano disruption. Theguardian.com
Tom Linton BV (2020) Coronovirus is proving we need more resilient supply chain. HBR.org
Underwood S (2016) Blockchain beyond bitcoin. Commun ACM 59(11):15–17
Xu L, Chen L, Gao Z, Chang Y, Iakovou E, Shi W (2018) Binding the physical and cyber worlds: a blockchain approach for cargo supply chain security enhancement. In: IEEE international symposium on technologies for homeland security (HST), pp 1–5
Yli-Huumo J, Ko D, Choi S, Park S, Smolander K (2016) Where is current research on blockchain technology? A system-atic review. PLoS One 1110

A Qualitative Study on Supply Chain Risk Management Adopting Blockchain Technology

Arpit Singh, Ashish Dwivedi, and Dindayal Agrawal

1 Introduction

Blockchain Technology (BCT) posits as a means of transferring a unit of value reliably without the inclusion of a financial intermediary such as banks (Dutta et al. 2020). BCT term emerged in the middle of the financial crisis. Primarily, BCT was a concept, dominant in the financial world pertaining to a new form of cryptocurrency called Bitcoin (Kouhizadeh et al. 2021). Many other cryptocurrencies such as Ethereum, Litecoin, and Dogecoin followed after bitcoin (Tönnissen and Teuteberg 2020). However, the applicability of BCT extended far beyond the financial world, and some creative applications of the technology have surfaced in the domain of Supply Chain Management (SCM) (Perera et al. 2020).

BCT offers numerous benefits in SCM. The most important ones include minimizing compliance risks such as corruption, fraud, export controls and sanctions, and labor law compliance. Some of the common problems associated with supply chain activities especially pertaining to lack of proper visibility like product origin, transformation, and movement can be addressed efficiently by adopting BCT (Janssen et al. 2020). Many legal and regulatory requirements around sanctions, customs, sourcing, etc. can be dealt by product tracking in supply chain (Dwivedi et al. 2021; Pan et al. 2020).

SCM encompasses a host of activities that rely heavily on real-time information for proper execution (Bai and Sarkis 2020). However, there have been challenges to collect real-time information for supply chains (Ali et al. 2021). With the increasing

A. Singh · A. Dwivedi (✉)
Jindal Global Business School, Jindal Global University, Sonipat, O.P, India
e-mail: ashish0852@gmail.com

A. Singh
e-mail: asingh6@jgu.edu.in

D. Agrawal
SOIL School of Business Design, Gurugram, Haryana, India

S. K. Paul et al. (eds.), *Supply Chain Risk and Disruption Management*, Flexible Systems Management, https://doi.org/10.1007/978-981-99-2629-9_7

complexity and geographical distribution of supply chains in the perpetual growing global economy, the problem of the information gap gets even more pronounced (Lim et al. 2021). Some digital technologies such as Enterprise Resource Planning (ERP) have shown potential in transforming the availability of real-time information to supply chain managers. The advantages offered by ERP are often outweighed by issues such as centralized nature, slow speed, and expensive.

The present study attempts to analyze the foundations of BCT and the possibilities of digital technology in managing logistics and supply chain risks (Dwivedi et al. 2022). The pivotal features of BCT that promise risk management associated with supply chains are transparency, validation, automation, and tokenization (Garg et al. 2021). Supply Chain Risk Management (SCRM) refers to identifying and devising strategies categorically to control and mitigate risks associated with the supply chains. Many retailers have taken significant steps in managing supply chains with the adoption of BCT. In the year 2018, when a series of food-borne illnesses cases emerged particularly E.Coli in lettuce, Salmonella in eggs, etc. Walmart suggested that suppliers of green leafy vegetables incorporate BCT to track down the products to the farm. The retailer continued working with International Business Machines (IBM) to implement BCT in order to maintain food safety (Sharma and Kumar 2021). Also, the French retail giant Carrefour group announced to incorporate BCT with the assistance from IBM to track the supply chains of chicken, tomatoes, and other food items. The plan of the Carrefour group was to eventually deploy the technology for the supply chain of all the fresh food items (Botelho et al. 2020). Walmart stated that traditional paper-based ledgers would take almost seven days to two weeks to track down the product's origin but with BCT it would be a matter of just about a few seconds to locate the source of contamination. With this technology, consumers can scan the product codes and trace the origin of the product almost instantly. The provider of the world's 65% of food grain and 70% of chocolate, Buhler group, is also implementing BCT to provide customers a greater degree of transparency about the quality of the food grain (Microsoft Industry Blogs 2022).

Proponents of BCT have acknowledged it as the technology with innumerable benefits and tremendous potential. Initially used for cryptocurrency, blockchains are progressively added and cryptographically recorded in digital ledgers after verification by each participant. The chain becomes increasingly rigid and immutable as more participants are added (Alfa et al. 2021).

The transparency and real-time information sharing attribute of BCT are particularly useful in supply chains due to various risks involved such as contamination, counterfeit components, missing documentation, shipping delays, and missed payments (Pratap et al. 2021). The above-mentioned risks not only result in cost overruns but also ruin and damage the brand. The research questions for the study are as follows:

- How can BCT impact SCM and bring it towards effective risk management?
- How can information sharing be achieved with BCT adoption in SCM?
- How can BCT be implemented in the supply chain of companies today?

2 Literature Review

2.1 Supply Chain Management

SCM broadly implies the management of all the activities and processes involved in the transformation of raw materials into final goods (Wieland 2021). It involves streamlining the supply part activities to maximize the customer value and attain a competitive edge in the market (Sodhi and Tang 2021). SCM indicates the efforts made by the suppliers to develop and maintain the supply chains that are efficient and cost-effective (Zekhnini et al. 2020). Supply chain encompasses a range of processes from production and product maintenance to the information systems that are required to execute the processes (Birkel and Müller 2021). SCM is essentially a centralized management phenomenon that controls the production, shipment, and distribution of the product (Esmaeilian et al. 2020). The effective management of supply chains can assist in reducing costs and increase customer satisfaction through faster product delivery (Wamba and Queiroz 2020). This can be achieved responsibly and strictly by controlling the internal inventories, keeping track of the internal production and sales, and ensuring the smooth distribution of goods and services. Every product or a unit of service that reaches the market is a collective effort of all the participants in a supply chain. This collective effort ensures the timely availability and quality of the finished product or service (Del Giudice et al. 2020). A supply chain manager typically coordinates all the activities of the supply chain that consists of planning, identifying the source of raw materials, maintaining efficient and productive manufacturing process, delivery and logistics, and the return process for the unwanted or damaged items. SCM helps the organizations achieve several business objectives. For instance, controlling manufacturing processes can lead to a better quality product with minimum cost overrun incurred due to shortages and inventory over supply.

2.2 Supply Chain Risk Management (SCRM)

SCRM is the sequence of steps to identify, assess, control, and mitigate the risks in the supply chain (Abdel-Basset and Mohamed 2020). A comprehensive approach to SCRM involves the management of risks at all levels of the supply chain for all the risk objects including suppliers, locations, and more. The three major stages comprising SCRM are risk identification where a risk profile is established and active monitoring is done to update it. Once the risks are identified, the areas with maximum impact are judged, and finally, proper active and reactive action plans are formulated to secure supply chain (El Baz and Ruel 2021). The various risks confronting supply chains are Financial risks that originate when suppliers suspect a threat to their financial health that can happen due to supplier's bankruptcy, market volatility, etc. Reputational risk occurs when the reputation of the supply chain is in danger which is under Corporate

Social Responsibility (CSR) and compliance issues. Natural disaster risk destroys supply chains due to natural disasters such as earthquakes, floods, or other natural hazards. Other risks include man-made risks, geopolitical risks where supply chain is disrupted due to global political events, and cyber risks that are posed due to the technology used by the suppliers (Xu et al. 2020).

2.3 Blockchain Technology (BCT)

BCT is an Internet-based technology that is valued for its exceptional ability to validate the transactions publicly and record them in immutable encrypted ledgers (Dutta et al. 2020). The technology was initially employed to store bitcoin transactions, which is a digital cryptocurrency. Essentially, Blockchain provided a platform for maintaining ledger and record of bitcoin transactions to millions of computers linked to networks across the globe. When two parties enter into a transaction, they are assigned keys that are cryptographically secured (Patil et al. 2020). This encrypted transaction creates a new block and is then validated by the blockchain (Ali et al. 2020). This new block is appended to the chain. The blockchain transaction gets completed with the block added to the chain and the ledger is updated (Patil et al. 2021). There are broadly two kinds of blockchain: "Permissionless" distributed ledgers such as bitcoin which are in the public domain and "Permissioned" ledgers that are largely centralized and controlled by agents like "actors", "nodes", and "miners" (Peng et al. 2021).

Although there is an overall positive sentiment for the adoption of BCT in the SCRM, the research on the topic is relatively scant. The present study attempts to shed light on the potential adoption of BCT in supply chains through interviews conducted with the experts in the relevant field.

3 Research Methodology

3.1 Research Design

The research is based on Interpretivism theory which refers to the process of deriving knowledge by leaning on the interpretation of meanings that humans associate with their actions (Walsham 1995). The study falls broadly in the qualitative research category. The reason for using qualitative methods to conduct the research was primarily to extract the in-depth understanding of the propensity of organizations including industries and academia for using BCT for effective risk management. The perception of the concerned population about the knowledge of BCT and its utility in SCRM was studied by conducting in-depth interviewees with the experts in their respective fields. The obtained qualitative data was subjected to interpretations and conclusions were drawn for the research objectives stated for the study.

3.2 Sampling and Data Collection

The sampling for the study was based on judgemental sampling in conjunction with snowball sampling. The sample included academicians actively working in the field of Information Technology (IT), informatics, and risk management. There were four respondents selected for the study having the basic understanding of various new technologies including BCT and were employed with the organizations for more than 5 years. Two respondents were the college instructor and research scholar that were working in the domain of IT and risk management. The semi-structured interviews with each respondent was done for around thirty to forty five minutes with further probing by interviewer as the requirement or necessity arose. The interviews were done online using the Zoom platform. Although the initial length of the interview was kept at 45 min maximum, some of them took more than one hour. The respondents requested to stay anonymous, so we denote each respondent by serial numbers 1–5.

3.3 Data Analysis

The recorded video sessions were analyzed thoroughly for each question asked from each respondent. The feature of transcribing the video session was immensely helpful. It worked as a ready resource to understand clearly without missing any significant point raised by the respondents. At first, some respondents expressed confusion in understanding the utility of BCT in SCM. They presumed that BCT is synonymous to bitcoins and it is useful only in the context of finance. However, they were informed about the potential of BCT in restructuring business models in various industries. They brainstormed ideas on how BCT are helpful in SCM and finally agreed to share their opinions.

4 Findings

4.1 Interviewee 1

RQ1: How can BCT be used for effective risk management in SCM?

Response: Based on my basic knowledge of blockchain, in terms of business, blockchain keeps a ledger of all the transactions that are involved throughout the supply chain in the form of blocks. To understand how it brings SCM towards effective risk management, I will use the simple example of the diamond industry. Diamond is a precious stone which holds value only if it is real. As soon as the diamond leaves the mined location and reaches another warehouse, this transaction will be recorded on the blockchain and the same when it is brought from the store. Now, if the customer wants to check its authenticity, he can take a look at the blockchain, where he will get all the information regarding the origin of the diamond, date, price, certification, and other things. This will guarantee the authenticity of the diamond. Having implemented blockchain in SCM can adversely reduce the risk of fraud or counterfeiting and creates transparency about the product or transaction for everyone involved in the supply chain.

RQ2: How can information sharing be achieved with BCT adoption in supply chain management?

Response: I will answer both your questions with this (research questions RQ1 and RQ2). First, we need to understand that each supply chain is unique in itself so to say how effectively it can help in managing the risk depends entirely on the supply chain and the type of smart contract between the parties. But if we talk in general, what blockchain does is it consolidates all the stakeholders in the supply chain, with the purpose of giving equal weightage to all the stakeholders in the supply chain. This helps keep track of all stakeholders in the supply chain, facilitate information sharing, and create transparency even in long supply chains. When talking about long supply chain, let us talk about extended supply chain, where the chain does not end when the product reaches the customer, it extends to after-services. For example, when you buy a car the supply chain does not end when you bought it. It extends to after-services of maintenance, which is extended supply chain of the previous one. Blockchain acts as a great facilitator in such extended supply chains by keeping a record of different transactions and future insight regarding after-services.

RQ3: How can BCT be implemented in the supply chain of companies today?

Response: Understanding how it can be implemented totally depends on what kind of supply chain and the company we are dealing with. "Smart Contracts" play a crucial part in the implementation of blockchain in companies. Smart contract is like a digital contract which is based on how much two parties involved in the transactions are willing to share information over the blockchain. This is a key factor that companies should keep in mind to efficiently set up blockchain in their supply chain. To understand it better you can take the example of Dubai, where the government is planning to record all of its government transactions over blockchain. There are a very small number of companies that have implemented blockchain in their supply chain and one of the reasons for the same can be the lack of knowledge about it if we do not take into consideration the cryptomarket that is totally different.

4.2 Interviewee 2

RQ1: How can BCT be used for effective risk management in supply chain management?

Response: The second interviewee reiterated the importance of BCT in effective risk management in supply chain through the example of Walmart. The interviewee stated that because the transactions recorded in a blockchain are traceable and immutable, it is tamper-proof. The product taken from destination A to destination B can be easily tracked for quality as well as timely delivery with the help of timestamps. Further, he added that he was closely observing how the BCT be applied in the construction sector to avoid accidents and other occupational hazards occurring at the workplace.

RQ2: How can information sharing be achieved with BCT in SCM?

Response: The interviewee took the example of a case study that he worked on during his research that involved the transportation of food containers from a food company to ABC airlines (anonymous name to ensure confidentiality). Initially, the order placed was for 100 containers of food but the actual delivery was 80. He argued that implementing blockchain in logistics would have helped in finding out where the 20 containers were, if delivered. Information sharing is a feature that is strongly advocated in Blockchain paradigm. Further, two blockchains can be interlinked to encourage cross-information sharing which the stakeholders can hugely benefit from.

RQ3: How can BCT be implemented in the supply chain of companies today?

Response: The interviewee answered this question from the experience as a researcher in the field of new technologies in the industrial sector. He said that not every industry requires the use of blockchain. Particularly, the small-scale sector where already some specific softwares are in place and functioning as desired. He adds that Blockchain has been considered as a "must-have" just because everyone is using it. Blockchain is just a software that should be used wherever it is needed. The property of Blockchain to make the records tamper-proof and immutable should be used to track the timely and safe delivery of products. Since a proper record of the transaction can be safely documented in the Blockchain, it can be used to address after-sale issues.

4.3 Interviewee 3

RQ1: How can BCT impact SCM and bring it towards effective risk management?

Response: The third interviewee explained that the major risks involved in the supply chain are the risks due to fraudulent activities and operational risks. She said that in a blockchain all the blocks or nodes contain the same data. So, if any tampering is done with the data then all the nodes have to be altered for the change. This is practically impossible since there are many blocks in a typical blockchain which are governed by various users. Thus, the probability of fraud in a supply chain gets significantly low.

BCT renders a paperless mode of working where all the transactions are recorded and stored securely. Internet of Things (IoT) and other smart devices should be adapted to complex supply chain operations. BCT is majorly applied in production and logistics but not so much in distribution.

RQ2: How can information sharing be achieved with Blockchain technology in supply chain management?

Response: As per the third interviewee, information sharing is facilitated by the decentralization feature of the blockchain. As the same information is on each block of the chain, anybody at any given time will have complete access to the information about the supply chain activities. The interviewee clarifies that having access to the information does not imply sharing confidential information. There are some facts and figures that are not disclosed on a public platform. For this, blockchain uses cryptographic feature through public and private keys that ensure confidential information is not out in public. Interviewee insisted strongly that organizations should check their requirements and capacities thoroughly before implementing blockchain in their organization.

Q3: How can BCT be implemented in the supply chain of companies today?

Response: The third interviewee mentioned that there are five major areas in SCM. The first one is financing followed by production, warehousing, logistics, and distribution/trading activities. The effective management of supply chain requires transparency and operational efficiency. The immutability and transparency attributes of BCT lend itself for a better management of supply chain which suffers from inefficiency due to operational disturbances and occasionally due to fraudulent activities.

This means that once a record has been added in a blockchain, then it is impossible to change it. Thus, if there is a possibility of corruption in the organization leading to tampering the important records and transactions, it is scrutinized by the blockchain itself that comprises a number of participants of the supply chain. Also, the third interviewee added that it is very important to initiate the blockchain with the correct information otherwise if a change of information is required at a later stage it would mean to start the blockchain over again. There is no constraint in adopting Blockchain framework in terms of company size but proper requirement and budget analysis should be done in order to implement the technology in any organization.

4.4 Interviewee 4

RQ1: How can Blockchain technology be used for effective risk management in SCM?

Response: The fourth interviewee reiterated the immutability and transparency of Blockchain being the pivotal pillars in ensuring effective risk management in supply chains. Since the records once entered cannot be altered in the blockchain setup, the information pertaining to the supply chains including managing the delivery of products, tracing the defects, and attending to customer enquiries after sales are easily managed. The interviewee again cited Walmart as an organization that has leveraged BCT for tracing the cause of the delivered spoiled food item in its supply chain.

RQ2: How can information sharing be achieved with BCT in supply chain management?

Response: The interviewee cited the immutability, transparency, and traceability features of blockchain as the key ingredient to establish trust in the supply chain. Since the data entered is verified by all the participants before it becomes a block in the chain, it is impossible to amend, delete, or edit the data without intimating the corresponding participant. The current supply chains are not completely traceable due to the involvement of multiple partners and there is a lack of immutability as the partners involved in a supply chain can make changes that are not easily tracked. Further, the interviewee adds that blockchain being a database is not error-free. The error can creep into the system if a wrong record is added at the onset of blockchain. The wrong record keeps on propagating through the entire system until it corrupts the blockchain. Thus, it is paramount to have as little human intervention as possible and resort to IoT-based technology in conjunction with blockchain. This would lead to have a validated data entry into the system.

RQ3: How can BCT be implemented in the supply chain of companies today?

Response: The interviewee reiterated that blockchain is essentially a new form of data structure. The principle of achieving transparent and trustless transaction in bitcoin is what distinguishes blockchain from other databases. The major issue plaguing supply chains today is the lack of trust. Ironically, blockchain builds trust by being trustless. This suggests the basic premise on which BCT is based in the organization's understanding of what they do well and what is it that others find attractive to do transactions with them. This situation invariably maintains the trust of the participants on blockchain-based transactions. Interviewee mentioned that a standalone blockchain is not sufficient in maintaining the authenticity of records entered into a supply chain. It has to be coupled with intelligent devices that perform accurate data entry operations in blockchain with little human intervention.

5 Managerial Implications

The extant literature in the context of BCT adoption in supply chain risk management has highlighted the importance of integrating blockchain technology with the existing business processes with the collaboration of various stakeholders involved in an organization. BCT is found to efficiently address the issues related to the volatility, security, uncertainty, and complexity in the supply chain processes. With careful measures taken to inform employees and stakeholders about the BCT and its usage, the practitioners can effectively implement BCT in the organizational processes. The management and practitioners can leverage the findings of the study to highlight the value added by BCT in the operational and risk management activities that will bolster the implementation of BCT with ease. Some of the important implications for the managers are described as follows:

- There is an urgent requirement to provide the knowledge and the necessary know-how to the employees in the organization about the importance and usage of BCT in the risk management system. Organizations should collaborate with institutions that assist in disseminating information about BCT and its applications. Seminars, conferences, and webinars can be used as effective medium to exchange information and address issues pertaining to the adoption of BCT in supply chain management.
- Organizations should foster greater engagement of all the stakeholders involved in the supply chain ecosystem with external agents including trading partners, government agencies, and regulators on various aspects of BCT. This will aid in developing a culture of sharing of ideas and knowledge leading to a better understanding of new technologies and their impact on the supply chain processes.
- Organizations that have not considered the adoption of BCT in the supply chain ecosystem will hopefully appreciate the utility offered by this technique in managing risks in supply chains. The senior management will understand and appreciate the efficacy brought about by the inclusion of BCT in supply chains thereby encouraging them to investigate and learn about this technology and its applications in risk management in supply chains. The study can prove to be a reality check for organizations who are keen to include BCT in their supply chains despite having limited resources. The technical feasibility and need are the two important determinants that organizations will have to comprehensively evaluate before deploying BCT-based risk management system in the supply chains.
- Finally, the organizations will come to a consensus on the usage of BCT in mitigating risks associated with the supply chain ecosystem. Blockchain operates on network effect which implies that more the number of people using it makes it more valuable. Enhanced collaboration among all the stakeholders in the organization will result in making the BCT as an efficient mechanism to counter risks in the supply chains.

6 Conclusions, Limitations, and Future Scope

The study aims to gain insights into how BCT be implemented in the supply chains to bring about effective risk management. The research involved an interpretive study where three Research Questions (RQs) were asked by the experts to gain an understanding on how does BCT lends itself towards managing risks associated with the supply chains. Four experts were interviewed on various aspects of the adoption of BCT in the supply chain management. The experts were academicians actively working in the field of Information Technology (IT), informatics, and risk management. The in-depth interviews were useful in eliciting key points pertaining to the management's perception towards adopting BCT in supply chain and major bottlenecks in the implementation of BCT-based operational framework in SCM.

Besides conducting interviews, an extensive literature review was performed to learn about the concept of supply chain and major risks in a supply chain and finally how BCT can be used to mitigate those risks.

In all the interviews conducted, the main themes associated with BCT namely immutability, transparency, and traceability were found to converge. The interviewees believed that BCT is in its infancy and should be given an opportunity to restructure the way supply chain operates. The study discovered that there is little awareness about this technology which impedes its adoption in the supply chain processes. It was observed that blockchain is often considered synonymous to bitcoins and that it is just a buzzword or a hype with no real tangible benefit associated with it. The interviews highlighted the importance of blockchain-based operating frameworks as opposed to legacy frameworks which are costlier to maintain. Almost all the interviewees agreed that Blockchain should not be used as a standalone software as any erroneous data entered in the first stage will keep propagating through the entire chain thus corrupting the system. It was suggested that including smart IoT-based equipment to input the data in a blockchain with little human intervention will eliminate the issues pertaining to corrupt data.

To summarize the outcome of the interviews, it can be asserted that a critical assessment is required to understand the organization's needs. Once the requirements are understood, a detailed understanding of BCT should be done. It was well established by the interviewees that most of the organizations want to adopt blockchain because it is a "hot software". This is misleading as each software has its own requirement, audience, raw materials, and skill sets. Traceability and digitalization features provided by BCT will increase operational efficiency by going paperless. Also, Blockchain coupled with Artificial Intelligence (AI) and IoT-based devices will help prevent fraud due to human intervention. This current exploratory study mainly focuses on highlighting the blockchain as a tool to enhance the effective risk management in the supply chain. The study helps management to understand the technology better guide them to implement BCT to manage risks in the supply chain.

Additionally, the study attempts to identify critical issues associated with the adoption of BCT in SCM especially in the case of a developing economy. It was noted that lack of proper awareness and some misinformation regarding the usage of new technology impedes management to implement BCT-based framework in organizations. It is of paramount importance to conduct awareness programs that sensitize the management of organizations to the concept of blockchain and its utility in risk management.

The theme of blockchain and its application in risk management is still to be explored extensively. The study presented here anticipates that the concerned management will consider this technology as a potential candidate for resolving various risks associated with SCM. Due to time and availability constraints, only four experts were interviewed for collecting insights on the adoption of BCT in risk management. Also, the study is qualitative in nature that does not analyze the problem of BCT adoption using data collected from surveys. Since the crux of the study is the opinion of the experts, there is ample opportunity to bring in more number of experts that can provide multi-dimensional analysis of the situation. The profile of the experts is limited to academicians working in the field of supply chain risk management and IT. We shall reach out to more experts from various fields and research domains and conduct more detailed and in-depth interviews to achieve a holistic understanding of how well blockchain is placed in the current industrial sector and how can it be used for effective SCM. In addition, the interviews were conducted at one point in time which might have some shortcomings in terms of getting a full-fledged and a longitudinal idea of how the mindset and perception of experts vary over time about the adoption of BCT. We propose to conduct a longer time horizon study to address the aforementioned caveat.

Conflict of Interest The authors declare that there are no conflicts of interest.

References

Abdel-Basset M, Mohamed R (2020) A novel plithogenic TOPSIS-CRITIC model for sustainable supply chain risk management. J Clean Prod 247:119586

Alfa AA, Alhassan JK, Olaniyi OM, Olalere M (2021) Blockchain technology in IoT systems: current trends, methodology, problems, applications, and future directions. J Reliab Intell Environ 7(2):115–143

Ali O, Ally M, Dwivedi Y (2020) The state of play of blockchain technology in the financial services sector: a systematic literature review. Int J Inf Manag 54:102199

Ali O, Jaradat A, Kulakli A, Abuhalimeh A (2021) A comparative study: blockchain technology utilization benefits, challenges and functionalities. IEEE Access 9:12730–12749

Bai C, Sarkis J (2020) A supply chain transparency and sustainability technology appraisal model for blockchain technology. Int J Prod Res 58(7):2142–2162

Birkel H, Müller JM (2021) Potentials of industry 4.0 for supply chain management within the triple bottom line of sustainability—A systematic literature review. J Clean Prod 289:125612

Botelho A, Silva IR, Ribeiro L, Lopes MS, Au-Yong-Oliveira M (2020) Improving food transparency through innovation and blockchain technology. In: European conference on innovation and entrepreneurship. Academic Conferences International Limited., pp 128–136

Del Giudice M, Chierici R, Mazzucchelli A, Fiano F (2020) Supply chain management in the era of circular economy: the moderating effect of big data. Int J Logist Manag 32(2):337–356

Dutta P, Choi TM, Somani S, Butala R (2020) Blockchain technology in supply chain operations: applications, challenges and research opportunities. Transp Res Part E: Logist Transp Rev 142:102067

Dwivedi A, Agrawal D, Paul SK, Pratap S (2022) Modeling the blockchain readiness challenges for product recovery system. Ann Oper Res 1–45

Dwivedi A, Agrawal D, Jha A, Gastaldi M, Paul SK, D'Adamo I (2021) Addressing the challenges to sustainable initiatives in value chain flexibility: implications for sustainable development goals. Glob J Flex Syst Manag 22(2):179–197

El Baz J, Ruel S (2021) Can supply chain risk management practices mitigate the disruption impacts on supply chains' resilience and robustness? Evidence from an empirical survey in a COVID-19 outbreak era. Int J Prod Econ 233:107972

Esmaeilian B, Sarkis J, Lewis K, Behdad S (2020) Blockchain for the future of sustainable supply chain management in industry 4.0. Resour Conserv Recycl 163:105064

Garg P, Gupta B, Chauhan AK, Sivarajah U, Gupta S, Modgil S (2021) Measuring the perceived benefits of implementing blockchain technology in the banking sector. Technol Forecast Soc Chang 163:120407

Janssen M, Weerakkody V, Ismagilova E, Sivarajah U, Irani Z (2020) A framework for analysing blockchain technology adoption: integrating institutional, market and technical factors. Int J Inf Manag 50:302–309

Kouhizadeh M, Saberi S, Sarkis J (2021) Blockchain technology and the sustainable supply chain: theoretically exploring adoption barriers. Int J Prod Econ 231:107831

Lim MK, Li Y, Wang C, Tseng ML (2021) A literature review of blockchain technology applications in supply chains: a comprehensive analysis of themes, methodologies and industries. Comput Ind Eng 154:107133

Microsoft Industry Blogs. https://cloudblogs.microsoft.com/industry-blog/manufacturing/2018/09/25/buhler-will-track-crops-from-farm-to-fork-using-blockchain-technology/. Accessed 15 Mar 2022

Pan X, Pan X, Song M, Ai B, Ming Y (2020) Blockchain technology and enterprise operational capabilities: an empirical test. Int J Inf Manag 52:101946

Patil A, Shardeo V, Dwivedi A, Madaan J (2020) An integrated approach to model the blockchain implementation barriers in humanitarian supply chain. J Glob Oper Strateg Sour 14(1):81–103

Patil A, Shardeo V, Dwivedi A, Madaan J (2021) Humanitarian logistics performance improvement model using blockchain approach. In: 11th annual international conference on industrial engineering and operations management Singapore, pp 1002–1013

Peng L, Feng W, Yan Z, Li Y, Zhou X, Shimizu S (2021) Privacy preservation in permissionless blockchain: a survey. Digit Commun Netw 7(3):295–307

Perera S, Nanayakkara S, Rodrigo MNN, Senaratne S, Weinand R (2020) Blockchain technology: is it hype or real in the construction industry? J Ind Inf Integr 17:100125

Pratap S, Daultani Y, Dwivedi A, Zhou F (2021) Supplier selection and evaluation in e-commerce enterprises: a data envelopment analysis approach. Benchmark: Int J 29(1):325–341

Sharma M, Kumar P (2021) Adoption of blockchain technology: a case study of walmart. In: Blockchain technology and applications for digital marketing. IGI Global, pp 210–225

Sodhi MS, Tang CS (2021) Supply chain management for extreme conditions: research opportunities. J Supply Chain Manag 57(1):7–16

Tönnissen S, Teuteberg F (2020) Analysing the impact of blockchain-technology for operations and supply chain management: an explanatory model drawn from multiple case studies. Int J Inf Manag 52:101953

Walsham G (1995) The emergence of interpretivism in IS research. Inf Syst Res 6(4):376–394

Wamba SF, Queiroz MM (2020) Blockchain in the operations and supply chain management: benefits, challenges and future research opportunities. Int J Inf Manag 52:102064

Wieland A (2021) Dancing the supply chain: toward transformative supply chain management. J Supply Chain Manag 57(1):58–73

Xu S, Zhang X, Feng L, Yang W (2020) Disruption risks in supply chain management: a literature review based on bibliometric analysis. Int J Prod Res 58(11):3508–3526

Zekhnini K, Cherrafi A, Bouhaddou I, Benghabrit Y, Garza-Reyes JA (2020) Supply chain management 4.0: a literature review and research framework. Benchmark: Int J 28(2):465–501

Assessment of Risks and Risk Management for Agriculture Supply Chain

Sneha Kumari, V. G. Venkatesh, and Yangyan Shi

1 Introduction

Agriculture has transformed from ancient traditional agriculture of the 1950s, mainly labor workforce-dependent, to modern smart agriculture. The studies reveal that the agricultural production processes have resulted in high agricultural production due to modern technology's contributions. The rise in agricultural productivity has led to the growth of factories and industries (Liu et al. 2020). The need for factories and the processing industry has helped generate revenue from fruits and vegetables and increased employment for the youth (Christiaensen et al. 2021). The growing agriculture productivity and initiatives have raised serious concerns for the agriculture supply chain (ASC), topical in the international markets. The agriculture supply chain from the field to factories and the processing industries has helped generate revenue from fruits and vegetables and increased employment for the youth (Christiaensen et al. 2020). To build an export-oriented production environment, marketing agencies plan awareness and promotional activities. In the meantime, stakeholders are keen to assess the risks and their management strategies through the institutional, climate, lead-time, and financial perspectives (Imbiri et al. 2021). The risks in ASC cannot be compared with the risks in other supply chains. An ASC is different from other supply chains as it is driven by various risks like institutional, financial, climate, and market risks. The ASC deals with agriculture commodities, which has

S. Kumari
Symbioisis School of Economics, Symbiosis International (Deemed University), Pune, India

V. G. Venkatesh
EM Normandie Business School, Metis Lab, Le Havre, France

Y. Shi (✉)
School of Economics, Jiangsu University of Technology, Changzhou, China
e-mail: peter.shi@mq.edu.au

Macquarie Business School, Macquarie University, Sydney, Australia

S. K. Paul et al. (eds.), *Supply Chain Risk and Disruption Management*, Flexible Systems Management, https://doi.org/10.1007/978-981-99-2629-9_8

a short shelf life. Despite short shelf life, agricultural commodities cannot maintain the quality specifications with climate change. There have been studies on ASC risks that majorly focus on climate change. Very few reports analyze the risk factors for ASC in detail by linking them to practice. Thus, there is a scope to investigate the shelf-life of agricultural commodities and work on the knowledge gap to minimize the ASC risks. The necessary awareness and promotion activities must be planned through marketing agencies to build an export-oriented production environment. The ASC also needs to look at the infrastructure facilities for primary processing, packaging, storage, etc. for example, agricultural commodities like strawberries have a short shelf life. Maintaining the quality of such products becomes a concern for the stakeholders. Perishable commodities like bananas, mango, strawberries, tomatoes, citrus, and guava have many risks. These commodities cannot be stored for a long and have a very short shelf life. The produce needs special care during the production to protect it from insects, pests, and disease. The distributors claim poor quality specifications with time (Salah et al. 2019). Agricultural commodities such as food grains require proper storage facilities to protect from moisture and spoilage from pests and insects.

The ASC for expensive agriculture commodities such as saffron needs special care as the commodities are driven by poor traceability resulting in poor quality (Salah et al. 2019). The stakeholders, especially developing countries, do not have a firm hold on the ASC. This results in losses and disruptions in the ASC. Inadequate knowledge of the advantages of digital agriculture, lack of system integration, ease of use of the application, language barriers, low access to farm sites, lack of technical motivation, lack of information, and lack of infrastructure are the challenges that farmers have been facing in the ASC (Mittal 2001). More risks need to be documented for the efficient management of ASC. Therefore, it is essential to document all ASC risks and their management concerning the farmers and literature. It is also essential to derive the risk management strategies from the assessed risks in the ASC.

Against this background, this chapter explores the following research questions: *What are the risks in the agriculture supply chain, and What are the risk management strategies for the same?* The chapter will result in risk management strategies that could help the readers to understand the scope for improving the existing ASC. The chapter with detailed documentation of different types of risks will help the stakeholders of ASC to manage the supply chain carefully. The deliberation is based on the literature and field interview analysis of different risk types and their management strategies highlighted in the secondary database and qualitative data from the progressive farmers. The chapter also documents the result of interviews conducted with the progressive farmers. Questions asked to the progressive farmers during the discussion were

- What are the risks in the agriculture supply chain?
- How did you manage the risks for the Agriculture supply chain?
- What were the challenges and their remedial measures?

- Do you receive any support from any scheme or government body concerning the agriculture supply chain?

ASC creates new development paradigms that warrant more interaction with institutional and environmental factors (Sørensen et al. 2010). ASC is compelled to save water, smart agriculture, high-quality specifications, productive growth, pollution-free agriculture, and economies of value. The production stage has to be integrated with ASC to meet stakeholders' expectations. It helps to improve crops production—per drop, per acre, per rupee. For this, digital tools and techniques have a significant role in managing any ASC. To be precise, state-of-the-art technologies such as machine learning, urban farming, hydroponics, aeroponics, aquaponics, blockchain, etc., are significant for creating a global market (Salah et al. 2019). Digitalization offers relevant solutions to agricultural production management. It has established a ground-breaking platform for sustainable agriculture, making ASC more competitive (Kumari and Patil 2019). ASC risks are now being managed with artificial intelligence tools (Kumari et al. 2018). This chapter will act as a catalyst to bridge the gap between theory and practice. The researchers will be benefitted from understanding the ASC in a better way.

The rest of the chapter is structured as follows: Sect. 2 provides an overview of risks in ASC. Section 3 discusses the risk assessment frameworks. Section 4 records the feedback from stakeholders' interviews. Section 5 offers a plausible risk management strategy for agricultural inputs. Section 6 reveals the changing paradigms. Section 7 deliberates on the role of supply chain financing, and finally, Sect. 8 concludes the chapter.

2 Risks in Agriculture Supply Chain

An ASC is a complex web of functions. Each function comprises risks such as lack of exposure to the farmers, promotion activities for building an export-oriented production environment, lack of infrastructure facilities for primary processing, packaging, storage, inadequate market linkages, weak market intelligence, and lack of training. The ASC in the chapter mainly focuses on agriculture commodities, dairy supply chain, floriculture, and fisheries. Each sector is dominated by risk factors, which can be visualized in Fig. 1.

The treemap roots down the risks in different sectors. Firstly, the floriculture sector has marketing, infrastructure, and cold chain risks (Mittal 2007). The flowers, if not stored at an optimum temperature, may add to waste (Roy 2015). It has been observed that a fall in demand for floral products, along with a lack of infrastructure for their storage, results in their damage (Hulme et al. 2018). To manage the ASC risks in floriculture, there is a need for the delegation of services for an integrated model of floriculture (Messner et al. 2021). The dairy sector is dominated by risks like cold chain and technology applications (Mor et al. 2018).

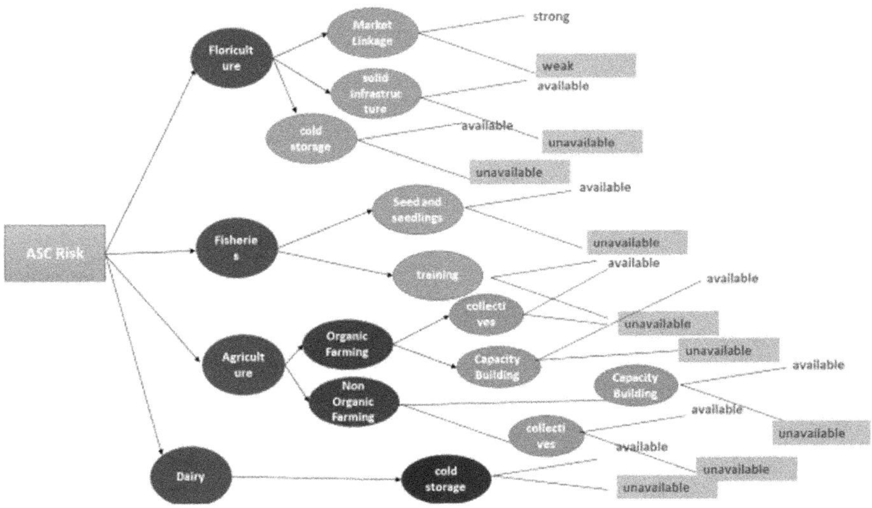

Fig. 1 Decision tree for primary issues for agriculture supply chain

The technology adoption for cold chains in the dairy sector helps increase income. The increase in milk procurement and processing capacity can create jobs for many unemployed and casual laborers. The animal husbandry department has fixed many subsidies for generating income and employment from poultry and farm. This sector, too, has a huge potential for income generation. Secondly, fisheries seeds and seedlings are often missing, which adds to the risk in this profession (Beck et al. 2011). There is a need for the conversion of stakeholders, the role of the bank, and training for income generation from fisheries. Thirdly, the agricultural sector has an untapped potential for exporting products that can be harnessed through 'natural farming.' There is a need for risk management in the agriculture sector through capacity building and collective action approach. There is a need to improve the shelf life of agricultural commodities and work on the knowledge gap. The investments in the cold chain have been triggered by market reform, investment subsidies, public service provision, and governance (Minten et al. 2014). Cold storage is associated with improved efficiency in supply chains and low waste (Kumari and Jeble 2020). Small farmers have the disadvantage of a lack of technical knowledge. Deficiency of technical skills and awareness about cold chain facilities are needed. Thus, Fig. 1 suggests the collective action approach, technology, and capacity building for agriculture risk management.

3 Risk Assessment of Agriculture Supply Chain Through Interviews with Progressive Farmers

This section explores the risks in ASCs through interviews with progressive farmers across the world. The 15 progressive farmers were approached through telephonic conversation for an interview on a purposive sampling approach, and the discussions were designed for 20–30 min addressing the questions on risks for ASCs. The insights from the key farmers are below.

Farmer 1—ASC depends upon the quality of agriculture production. Agriculture production is often challenged by inadequate rainfall and drought conditions. Transforming drought-hit farmland into a green belt is a prime challenge (Singh and Kumdhar 2016). The timely and effective water conservation approach coupled with soil, seed, and fertilizer management witnesses horticulture and pulses' quality bumper production, which brings a pleasant surprise to fellow farmers. Suitable applications of technology and knowledge in the farm field have not only made him a progressive farmer but took him to be an integral part of the governing body of a *World Bank-sponsored Agriculture Technology project*. He shifted into Agriculture diversification and marketing coordination. Technology application and knowledge are the keys to managing the risks. Most farmers face risks due to a lack of technology and unawareness. The prime element for ASC is the quality of agricultural produce. To meet the quality specifications, the farmers need to shift towards technology.

Farmer 2—Left the job and started the agriculture profession, even though people dissuaded him with the perception of no income from agriculture. He started working in Mango production. The land for the same was in the worst condition and passed on to them deliberately to discourage his agriculture venture. He worked consistently for 1 year and, with the support of scientists from the local leading agriculture university, adopted scientific techniques in pruning, cutting, and fertilizer applications. In a year, they increased the income from the mango field from twenty-five 350 USD to 1500 USD (roughly) within a year. This was an encouragement and an eye-opener for the farmers. Then his team started the production of Litchi with the help of drip irrigation and micro-sprinkler. They controlled the microclimate for litchi. They were in touch with processors, and for 8 days, the whole night, they harvested the litchi. This helped them increase their income from 1200 to 3500 USD (roughly). In 15 acres of land, the litchi today is sold at 40,000 USD annually. Agriculture has a huge potential with technology for increasing income. The team took the help of government subsidies to set up the complete automatic technology. They have good broadband and subscribed farm ERP system to digitize agriculture. He has been awarded many rewards for smart agriculture and online marketing. The turnover of his farm is roughly 100,000 USD, which will increase to 300,000 USD in the coming 5 years. While this is growing, they face marketing risks as the farmers should have good quality specifications for ASC. The farmers can increase productivity, but somewhere there is a need to look into the quality aspect of the agricultural produce. It is essential to focus on consumer-driven agriculture commodities. Consumer-driven products help in managing the market risks.

Farmer 3—One of the farmers was troubled by the barren lands. She ran a self-help group, and with its help, she built watersheds, tanks, check dams, and rainwater harvesting systems. All this helped her make the farm green and improve the water problems in the villages. Her initiatives towards making the villages green removed the water barriers for the crops. The farmers face the individual risk of having low landholdings and economies of scale. Therefore, it is essential for collective action by the farmers. The farmer also concluded that quality specification is the primary element in ASC. A collective or cluster-based approach needs to monitor and manage the quality. Farmers have shifted towards a collective approach for managing the risks of economies of scale.

Farmer 4—Attempted to store agricultural commodities such as wheat, coriander, mustard, and maize to avoid wastage. Therefore, there were restrictions, and they did not find access to the market for the ready crops. The farmers will face storage risks, but they need to learn to manage them. ASC risk is accompanied by the lack of infrastructure and inventory management issues, resulting in poor ASC performance. Risks like infrastructure and storage drive the ASC. Most of the agriculture commodities get wasted due to improper inventory management.

Farmer 5—The farmers grew maize, paddy, wheat, oats, and pulses. The major challenge was the procurement of agricultural inputs like seed, fertilizer, and tractors. This case tells us that agriculture has the potential to become viable. The need of the hour is to guide the farmers for effective pre-harvest and post-harvest management along with the proper market linkages. The farmers risk being untrained and having poor technical skills, leading to poor performance. The farmer pointed out a need for capacity building of the farmers and stakeholders. This can be done by providing proper training on producing and handling the agricultural produce.

The majority of the farmers agreed that inadequate rainfall, pests, insects, and disease affect the product's quality (Fig. 2). The poor infrastructure, lack of technology, storage structures, and unawareness are other risk factors in the ASC.

Fig. 2 Risks assessment based on the Farmers' Interview

4 Risk Assessment in Supply Chain of Agriculture Inputs Through Interviews

The farmers face risks in the supply chain of agriculture inputs. The risks involved with the agriculture inputs are explained below. The response from the farmers is clustered in Fig. 3.

Seed

Farmer 6—Good quality seed is a critical input to high productivity and farmers' welfare (Khanal and Maharjan 2015). The seed supply chain suffers from a weak regulatory mechanism, weak intellectual property rights (IPR) policy, and low investments in biotechnology. The risk management for the seed supply chain can be done through investment in R&D. The *Protection of Plant Varieties and Farmers' Rights Act* (PPVFRA), 2001 needs to be revisited to facilitate a robust IPR regime. Improving the financial position of the seed company, operation cost, logistics, and diverse agro-climatic conditions need to be carefully monitored for the supply chain.

Insecticide

Farmer 7—Losses caused by the pests have threatened the agricultural cropping system (Heong et al. 2015). The supply chain of insecticides depends on the information and the product, and the information supply chain needs to flow from the government agencies to farmers through training, media campaigns, extension activities, field study, and trials. Surveys report that most of the insecticides sprayed by the farmers are unnecessary and disrupt the supply chain of the agriculture inputs.

Fig. 3 Risks in agriculture supply chain for agri inputs

Fertilizers

Farmer 8—The fertilizer sector faces risk in the supply chain in the form of reformulation with minor hazardous components, storage, disposal, labeling, training, engineering controls, protective equipment, emission controls, incident monitoring, controls in raw material purchasing, and product sourcing, controls in storage, and distribution, auditing of operators, contractors, suppliers, responsible advertising, and after-sales support. Lack of knowledge about the usage of neem-coated urea (NCU), the challenge in calculating the recommended dose of fertilization (RDF) from different fertilizer brands, lack of knowledge about the method of fertilizer application, and lack of awareness about fertilizer use in crop farming are the major risks here. For these reasons, awareness amongst farmers regarding integrated nutrients management with NCU is required to be created. It can be done by more and more field demonstrations regarding the usage of NCU in the cultivation of crops and its use in other than crop production purposes, i.e., silage making, mixed with weedicide, and fisheries feed preparation to be conducted by a farmers field (Chouhan et al. 2018).

Drones

Farmer 9—The application of bar codes, radio frequency identification, and QR codes efficiently apply technology in agriculture and drone usage. The flight time and range vary in drones from 20 to 60 min, and drones are dependent upon the weather, knowledge, skills, and better connectivity (Tubis et al. 2021). However, the technology implementation requires awareness creation with user-training sessions, and the lack of exposure is the most significant risk in using the technology applications.

Farm Machinery

Farmer 10—The demand for farm machinery has increased over time. However, the risk in farm machinery is the accessibility and credibility of farm machines (Hinnou et al. 2022). Significant logistics issues such as order processing, inventory planning, distribution structure, and transportation issues contribute to the risks in the supply chain of tractors (Raghuram 2004).

Agrochemicals

Farmer 11—Smallholder farmers are not provided proper training and capacity building to use agrochemicals, leading to unsafe use of agrochemicals (Mengistie et al. 2016). Climate change and biodiversity norms have put restrictions on the agrochemical supply chain. Organic Compost Manure faces the challenge of a lack of infrastructure, legislation, and framework (Jiang et al. 2022).

5 Risk Management Strategies in Agriculture Supply Chain

ASCs are difficult to organize and stabilize in countries with many small and marginal farm holdings, and the production and aggregation parts must be efficient to achieve higher returns. The greatest job at hand is to encourage the small and marginal farmers and build confidence in them to move away from subsistence-level farming to market-oriented and remunerative agriculture through the adoption of newer technologies, post-harvest processing, and value additional agri-products at the community level. Considering several sub-activities in the ASCs, management of such sub-activities, financing at each level, and monitoring the supply chain activities through networking with stakeholders become problematic if one relies on financial assistance from Commercial banks. The community-level financial organizations in the form of collectives can meaningfully transform low-value primary agri-produce to higher values through demand-based contract farming, aggregating, processing, packaging, branding, transporting, warehousing, marketing/retailing, etc.

Collectives have the potential to transform agriculture into a profitable business venture through a well-coordinated collective action. The following are the possible intervention by which the ASC risks can be managed, as shown in Fig. 4.

- *Fruits and Vegetables Supply Chain*: Collectives are best suited for effective marketing of fruits and vegetables. Small and marginal farmer members can save themselves from being exploited by local traders during the flush season of production. In the absence of immediate local demand and facilities for transportation and storage, the collective effort can explore marketing potential in the nearby urban markets and fetch remunerative prices for their products (Poulton et al. 2010). Community-led fruit preservation or processing units can also process perishable

Fig. 4 Word cloud on risk management for agriculture supply chain

commodities, enhance shelf life, and realize more price than raw agri-produce. This will improve the confidence level of the growers and create additional jobs, particularly for the local small, marginal, and landless families.

- *Dairy Supply Chain*: Dairy is one of the most potent tools for ensuring sustainable livelihood income for millions of small farmers in rural India. Collectives promote dairy husbandry through efficient delivery of breeding and health care services at farmers' doorsteps. Collectives have the capabilities to ensure systematic effort to develop and operationalize the supply chain in the dairy and dairy processing sector. A little handholding by the government or an expert agency can ensure transparency, coordination, and networking of various dairy supply chain stakeholders to sustain the related business ventures.
- *Contract Farming*: Contract farming enables small and marginal farmers to participate in new high-value and diversified product markets and helps in improving quality standards to ensure remunerative prices for the products so produced by contract growers (Singh 2002). Since agri-markets are largely buyer-driven and vertically integrated, contract farming through community-based farmers would offer the best possible income stream to the farmers by reducing labor-related transaction costs, costs on technology and innovation, research and development, and their application in the field. In comparison to individual farmers, cooperative producer organizations can reap the benefits of lower input costs, stability, and longevity of contract farming arrangements and deliver a fair and sustainable distribution of profits amongst the member farmers. Co-operative producer organizations have the desired potential for balancing the complicated dynamics between firms and farmers through collective bargaining, creation, and maintenance of long-term relationships with input vendors and logistic support providers, and timely mitigating of risks and uncertainties faced by the farmers.
- *AgriInputs Supply Chain*: The Agri input market, both at the level of the user-farmer and the producer-investor, can be managed through the following measures: (i) farm-level extension and promotion programs, (ii) financial assistance to investors in setting up units, (iii) subsidies on sales, and (iv) direct production in the public sector and cooperative organizations and in universities and research institutions (Mazid and Khan 2015; Pal et al. 2015).
- *Co-operation andAgri-marketing*: Agri-marketing ensures a vital link between the farmers and consumers. Co-operative agriculture marketing has immense potential in resolving the complex and complicated problems faced by the present agri-marketing system. The strengthening and revival of the existing co-operative marketing system in the agriculture sector would eliminate not only excessive dependence of agents and intermediaries in the organized wholesale markets and unorganized rural periodical markets (Village Agricultural Markets) but also ensure appropriate price discovery by resolving issues of effective information dissemination, use of digitized means of marketing, management of transportation costs by joint transportation of commodities and establishment of a network of warehouses for storage of perishable and semi-perishable agri-commodities.

Setting up co-operative sale societies and co-operative warehousing units may be the best solutions to help the agriculturists realize the rightful profits on their output at the community level.

Small and marginal farmers' availability and access to markets are vital in designing the market infrastructure. At present, co-operative marketing consists of commission shops in various marketplaces. These shops neither undertake collection or aggregation of agri-produce from the farmers at the farms nor do they provide joint and cost-effective transportation and other logistic support for ensuring better price discovery. Further, the cooperative marketing system lacks adequate mechanisms to undertake and provide timely and adequate processing and preservation facilities for perishable/semi-perishable agriculture products. The need of the hour is to upgrade and strengthen the establishment of organized facility centers for aggregation and transportation of agri-commodities, assaying, pre-conditioning, grading, to standardize packaging and storage of the products. Thus, a solid and vibrant collective marketing infrastructure has significant potential in making ASC efficient by effective and timely dissemination of market intelligence and actual demand statistics of the commodity to be traded and the ruling prices of such items amongst the member farmers of the societies. Thus the need of the hour is to:

- Establish community-level hubs strategically placed in rural and urban growth centers.
- Ensure finance to such processing and value-addition units by assuring access to banking infrastructure or adequate and efficient public–private partnerships.
- Facilitate collectives and startups and encourage venture capitalists to invest in innovative agri-processing startups through appropriate policy interventions.
- Setting up adequate accredited food quality testing labs at convenient and strategic locations.
- Make available infrastructure for skill development and capacity building for farmer members to process and preserve perishable and semi-perishable agri-products.
- Impart training and essential orientation tips to members of collective marketing societies on grading, assaying, sorting, and standardization of agri-commodities.

The growth of the agriculture sector remains an important area of discussion for policymakers. In the present situation, the agriculture sector's significant constraints are controlling small and marginal farmers' financial and market conditions. The collective action of farmers can result in agriculture value addition and marketing (Levay 1983). Small farmers' critical concerns are inadequate extension services, low-level technology adoption, lack of capital, poor business skills, low income due to poor infrastructure, and low marketing efficiency. Many forms of aggregation in farmer interest groups, self-help groups, cooperatives, and *Farmer Producer Organizations* (FPOs) emerge as the most effective tools to manage the overall supply chain professionally. In addition to the challenges like ineffective leadership, small and marginal holdings of farmers, poor market linkages, inability to attract talent, absence of time-tested thinking and planning, ignorance of principles of basic accounting, and not knowing how to make a business plan for the organization, the farmers also lack understanding of the rules and regulations of a Company and the statutory requirement. There is also a need to collectively promote FPOs to handle multiple commodities for value addition and marketing. The ability of leaders' energy at the age of space-time can be a driver for sustainable agriculture growth.

6 Changing Dimensions for Risk Management in Agriculture Supply Chain

The agriculture supply chain needs to be integrated with technology and other dimensions, as shown in Fig. 5.

- Leadership—Professional leader who is willing to work for the farmers. Professionals and youths need to be attracted to collectives.
- Philosophy—The value for money and the value of many are essential for encouraging the farmers toward the supply chain. The incoming professionals are required for the supply chain despite the lucrative offers.
- Consistency—A sustained approach is required in communication and human resources planning; there is a need for a dedicated and efficient supply chain.
- Technology adoption at all levels—The low input and low output model needs to be changed to better information.
- Creation of dedicated supply chain—The supply chain demands a dedicated team in the supply chain who are consistent in their approach.
- Innovation—Regular education of the farmers is required for adaptation to innovation in collectives.
- Branding and Marketing is the essential element for a better supply chain.

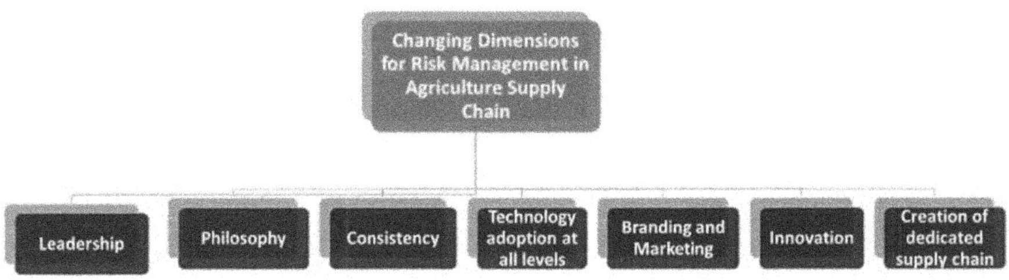

Fig. 5 Changing dimensions for risk management in agriculture supply chain

7 Agriculture Supply Chain Financing for Risk Management

In general, the farmers and stakeholders in emerging countries are not financially sound, and therefore tangible investments are one of the demanding needs for ASC. The financing requires substantial investments to adopt technology, infrastructure, or process. A significant effort exists for capacity building at various levels, i.e., farmers, managers, governing councils, and bankers. Limited access to better inputs, including credit and technologies, a low market surplus of food crops, and inadequate warehousing facilities, including cold storage, cold chain, or perishable, have made it very difficult for the ASC to sustain. ASC financing is required for capacity building, integrated management, IT-enabled technology services, infrastructure credit, biotechnology, processing, and aggregate financial models. Sustaining ASC financing in solid business and profitable lines is a challenging task because the sector is dominated by small landholders (Ceballos et al. 2020). In some cases, farmers are at a disadvantage due to scale limitations (Behzadi et al. 2018). To overcome the drawback, the rural population composed of cooperatives, regional banks, commercial banks, non-banking financial institutions, and lending agencies has supported agriculture through credit flow. ASC actors need finance for production, procurement, processing, storage, and distribution. The credit flow at each stage is essential for sustaining the ASC.

Financial institutions provide loans to the farmers based on their repayment and risk-bearing capacity. Small farmers have a meager compensation and risk-bearing ability, making it difficult for financial institutions to get credit flow. Financial agencies can indirectly engage the supply chain stakeholders to provide credit to the farmers. The economic approaches help in the value proposition, creation and delivery, customer relationships, capturing value, intentions, partnership, and collaboration. The financial institutions can follow the different economic approaches (Fig. 6).

- Indirect Supplier Financing: The financial agencies are aware that the farmers are not in a position to bear high risk. Therefore, they may support the ASC players, viz., agro-processors and market agencies who are more creditworthy and less prone to risk. The ASC players may take the risk of lending to the farmers to sustain ASC financing.
- Interdependence Financing: This approach has interdependent links. The financial agency is ready to provide credit to the farmers if they have the following link in the supply chain. The business success of one link is dependent upon the other connection.
- Cascade Financing: The financial agency targets the linkage of the supply chain. The agency may not finance the farmers in isolation but provides finance to the primary producers, processors, distributors, or end buyers.
- Joint Liability Group Financing: This approach has individual and group financing. Underfunding individuals, each member of the joint liability group

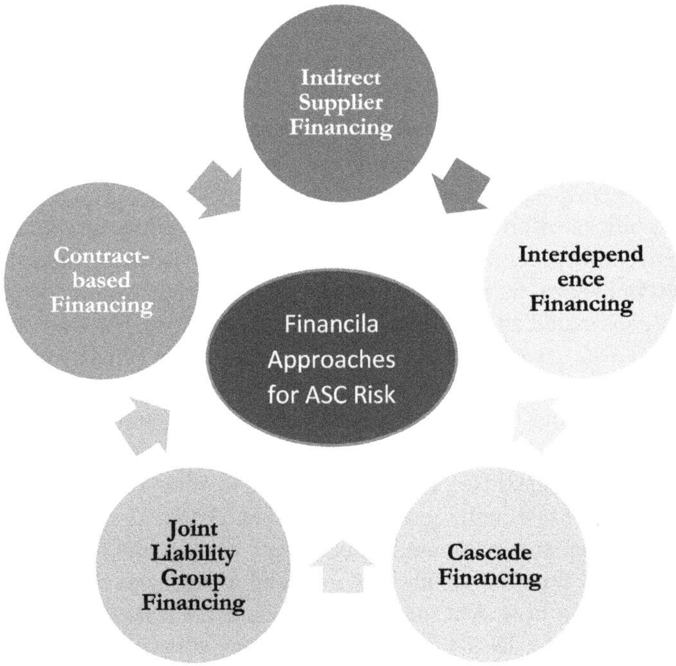

Fig. 6 Linking financial approaches for agri supply chain risk management

may be provided Kisan Credit Card (KCC). Under the group financing approach, the Joint liability Group acts as one borrowing unit.

- Contract-Based Financing: In this financial approach, the financial agency is willing to reduce the risk of the defaulters. The farmers who enter into a contract with some reliable buyers are financed, wherein the agreement ensures that the farmers' income may be used as loan repayments.

8 Implications

8.1 Theoretical Implications

Our study offers vast literature on ASC risk and risk management. The extensive literature review and qualitative approach add to the pool of knowledge on the ASC risks assessment and management. Overall, we adopted an interview-based approach for understanding the ASC risks and the strategies essential for managing the risks in ASC. Our findings also add to understanding the importance of risk management strategies. The study maps the practical challenges and risks in the ASC with the literature review.

8.2 *Practical Implications*

The findings have implications for policymakers, collectives, clusters, cooperatives, practitioners, and farmers. The schemes and policy initiatives promote uniform standards to integrate the stakeholders. It is time for the Government and Industry Association to work together to address the gaps in minimizing the ASC risks. Collective Action and network building will help strengthen the ASC in developing countries. The practitioners can understand the linkage of different managing tools for reducing the ASC risks. The policymakers need to understand the risk assessment to link the schemes with the ASC risk prioritization model. The study has more practical exposure than theoretical as the ASC needs to be managed by carefully working on the study results. The study has posed future research directions that can help the ASC practitioners.

9 Conclusions and Future Research Scopes

The study can help develop a road map for the youths to increase their income from the agriculture supply chain. Many very well-educated young persons with diverse fields have come into the agriculture supply chain, and ASC is coming up with a more innovative and upward-looking perspective. There is a need to derive solutions for developing connections between farmers and markets (Kumar et al. 2020a, b). The study is limited to an extensive literature survey and interviews conducted with the farmers. The result concludes that eco-innovation technology needs to be implemented in the supply chain to reduce risks (Hasler et al. 2016).

The domestic ASC is graduating towards global ASC demand. To make the agri sector a superpower, there is a need to bring suitable innovation for ASC financing. The three fundamental aspects of the agriculture supply chain are open market and transparency, agri-ecosystem, and farmers' perspective. The market should be more accessible to farmers, traders, entrepreneurs, and industry. Transparency policy should be welcomed because transparency will bring in more growth in ASC. Openness is the key factor in the development of the ASC. The 3S that require the constant attention of policymakers and the government are Suitability, Sustainability, and Scalability. Suitability provides exclusive benefits to farmers, stakeholders, entrepreneurs, and the industry. Sustainability focuses on the long-term vision of the agriculture sector. Scalability is the scale of numbers operating in the market, i.e., to make the system completely accessible. Agriculture is moving towards technology, where new technological trends have been adopted, like E-Commerce. ASC risk management strategies can help sustain the ASC.

The study is limited to secondary data and qualitative research. The findings are based on the interviews conducted with the farmers, and the results need to be validated by conducting a quantitative study. The study has come up with the critical antecedents for ASC risk management. The key result of the study is presented in

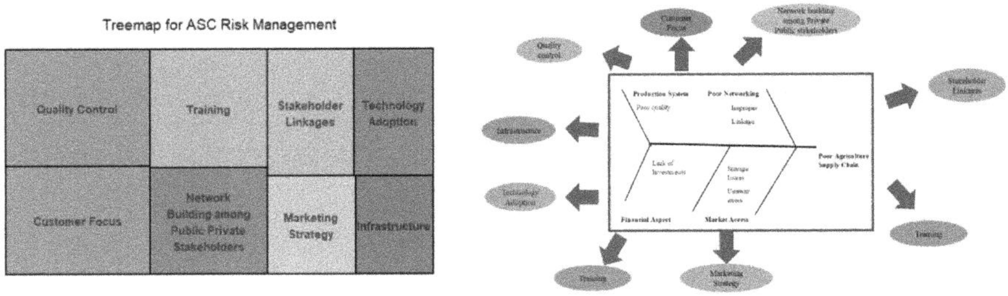

Fig. 7 Treemap and fish bone diagram for agriculture supply chain risk management

Fig. 7. The Fish Bone diagram explains the primary risks in ASC. The risks focus on improper linkages among the stakeholders, poor quality, poor network, lack of sources for investments, storage issues, and unawareness. Academicians and practitioners need to prioritize the risks. The risk prioritization has been further explained through treemap. The treemap has been constructed based on the farmers' responses regarding the risk managing strategies. The results show that quality specification and customer focus produce are highly preferred antecedents by the stakeholders.

The treemap and Fish Bone Diagram pose the below future research questions. These future research questions can help academicians and practitioners to reduce the complexity of ASC.

- How can the stakeholders maintain the quality of the agriculture commodities?
- What are the customer-focused products in current times?
- What kind of training is required for the farmers for ASC risk management?
- What are the drivers for network building among public–private partnerships?
- What are the marketing strategies for ASC?
- What are the drivers for technology adoption in ASC?
- How can the stakeholders be driven for technology adoption?

References

Beck MW, Brumbaugh RD, Airoldi L, Carranza A, Coen LD, Crawford C et al (2011) Oyster reefs at risk and recommendations for conservation, restoration, and management. Bioscience 61(2):107–116

Behzadi G, O'Sullivan MJ, Olsen TL, Zhang A (2018) Agribusiness supply chain risk management: a review of quantitative decision models. Omega 79:21–42

Ceballos F, Kannan S, Kramer B (2020) Impacts of a national lockdown on smallholder farmers' income and food security: empirical evidence from two states in India. World Dev 136:105069

Chouhan RS, Niranjan HK, Sharma HO, Rathi D, Kurmi HS (2018) Constraint in adoption of neem coated urea (NCU) in Madhya Pradesh. Int J Bio-Resour Stress Manag 9(1):173–177

Christiaensen L, Rutledge Z, Taylor JE (2020) The future of work in agriculture: some reflections. In: World Bank Policy Research Working Paper, no 9193

Christiaensen L, Rutledge Z, Taylor JE (2021) The future of work in agri-food. Food Policy 99:101963

Hasler K, Olfs HW, Omta O, Bröring S (2016) Drivers for the adoption of eco-innovations in the German fertilizer supply chain. Sustainability 8(8):682

Heong KL, Wong L, Delos Reyes JH (2015) Addressing planthopper threats to Asian rice farming and food security: fixing insecticide misuse. In: Rice planthoppers. Springer, Dordrecht, pp 65–76

Hinnou LC, Obossou EAR, Adjovi NRA (2022) Understanding the mechanisms of access and management of agricultural machinery in Benin. Sci Afr 15:e01121

Hulme PE, Brundu G, Carboni M, Dehnen-Schmutz K, Dullinger S, Early R et al (2018) Integrating invasive species policies across ornamental horticulture supply chains to prevent plant invasions. J Appl Ecol 55(1):92–98

Imbiri S, Rameezdeen R, Chileshe N, Statsenko L (2021) A novel taxonomy for risks in agribusiness supply chains: a systematic literature review. Sustainability 13(16):9217

Jiang Y, Li K, Chen S, Fu X, Feng S, Zhuang Z (2022) A sustainable agricultural supply chain considering substituting organic manure for chemical fertilizer. Sustain Prod Consum 29:432–446

Khanal NP, Maharjan KL (2015) Risk Management in community seed production under rice-Wheat cropping system. In: Community seed production sustainability in rice-wheat farming. Springer, Tokyo, pp 121–133

Kumar A, Padhee AK, Kumar S (2020a) How Indian agriculture should change after COVID-19. Food Secur 12(4):837–840

Kumar U, Raman RK, Kumar A, Singh DK, Mukherjee A, Singh J, Bhatt BP (2020b) Return migration of labours in bihar due to COVID-19: Status and strategies of deployment in agricultural sector. J Commun Mobil Sustain Dev 15(1):192–200

Kumari S, Patil YB (2019) Enablers of sustainable industrial ecosystem: framework and future research directions. Manag Environ Qual 30(1):61–86

Kumari S, Jeble S (2020) Waste management through industrial symbiosis: case study approach. Latin Am J Manag Sustain Dev 5(1):37–46

Kumari S, Jeble S, Patil YB (2018) Barriers to technology adoption in agriculture-based industry and its integration into technology acceptance model. Int J Agric Resour Gov Ecol 14(4):338–351

LeVay C (1983) Agricultural co-operative theory: a review. J Agric Econ 34(1):1–44

Liu Y, Ma X, Shu L, Hancke GP, Abu-Mahfouz AM (2020) From industry 4.0 to agriculture 4.0: current status, enabling technologies, and research challenges. IEEE Trans Ind Inform 17(6):4322–4334

Mazid M, Khan TA (2015) Future of bio-fertilizers in Indian agriculture: an overview. Int J Agricul Food Res 3(3)

Mengistie BT, Mol AP, Oosterveer P (2016) Private environmental governance in the Ethiopian pesticide supply chain: importation, distribution and use. NJAS-Wagening J Life Sci 76:65–73

Messner R, Johnson H, Richards C (2021) From surplus-to-waste: a study of systemic overproduction, surplus and food waste in horticultural supply chains. J Clean Prod 278:123952

Minten B, Reardon T, Singh KM, Sutradhar R (2014) The new and changing roles of cold storages in the potato supply chain in Bihar. Econ Polit Weekly 98–108

Mittal SC (2001) Role of information technology in agriculture and its scope in India. Fertil News 46(12):83–88

Mittal S (2007) Strengthening backward and forward linkages in horticulture: some successful initiatives. Agricul Econ Res Rev 20(347-2016-16832):457–469

Mor RS, Bhardwaj A, Singh S (2018) Benchmarking the interactions among performance indicators in dairy supply chain: an ISM approach. Benchmark: Int J 25(9):3858–3881

Pal S, Singh HB, Farooqui A, Rakshit A (2015) Fungal biofertilizers in Indian agriculture: perception, demand and promotion. J Eco-Friendly Agricul 10(2):101–113

Poulton C, Dorward A, Kydd J (2010) The future of small farms: new directions for services, institutions, and intermediation. World Dev 38(10):1413–1428

Raghuram G (2004) Logistics of tractor distribution in an agriculture-driven economy: an Indian case study. Int Trans Oper Res 11(6):701–714

Roy TN (2015) Supply chain management of horticultural crops. In: Value addition of horticultural crops: recent trends and future directions. Springer, New Delhi, pp 293–314
Salah K, Nizamuddin N, Jayaraman R, Omar M (2019) Blockchain-based soybean traceability in agricultural supply chain. IEEE Access 7:73295–73305
Singh S (2002) Contracting out solutions: Political economy of contract farming in the Indian Punjab. World Dev 30(9):1621–1638
Singh TP, Kumdhar VS (2016) Socio-economic status of farmers in drought prone region of Maharashtra, India-case study. Int J Curr Res 8(06):33304–33306
Sørensen CG, Fountas S, Nash E, Pesonen L, Bochtis D, Pedersen SM et al (2010) Conceptual model of a future farm management information system. Comput Electron Agricul 72(1):37–47
Tubis AA, Ryczyński J, Żurek A (2021) Risk assessment for the use of drones in warehouse operations in the first phase of introducing the service to the market. Sensors 21(20):6713

Resilience of Agri-Food Supply Chains: Australian Developments After a Decade of Supply and Demand Shocks

Firouzeh Rosa Taghikhah, Derek Baker, Moe Thander Wynn, Michael Billy Sung, Stuart Mounter, Michael Rosemann, and Alexey Voinov

1 Introduction

Australian agribusiness has made a significant transition in value addition since the 1970s. Examples include the red meat and wine industries, which have made transitions from commodities to premium products with strong export performance; and the expansion of supply to high-value domestic markets such as fish and fruits.

F. R. Taghikhah (✉)
Business School, University of Sydney, Sydney, Australia
e-mail: Firouzeh.th@gmail.com

D. Baker · S. Mounter
UNE Centre for Agribusiness, University of New England, Armidale, Australia
e-mail: abaker33@une.edu.au

S. Mounter
e-mail: smounte2@une.edu.au

M. T. Wynn
Faculty of Science and Engineering, Queensland University of Technology, Brisbane, Australia
e-mail: m.wynn@qut.edu.au

M. B. Sung
Consumer Research Lab, Curtin University, Perth, Australia
e-mail: billy.sung@curtin.edu.au

M. Rosemann
Faculty of Business & Law, School of Management, Queensland University of Technology, Brisbane, Australia
e-mail: m.rosemann@qut.edu.au

A. Voinov
University of Twente, Enschede, Netherlands
e-mail: aavoinov@gmail.com

F. R. Taghikhah · D. Baker · M. T. Wynn · M. B. Sung · S. Mounter · M. Rosemann · A. Voinov
Food Agility CRC Ltd., Sydney, NSW, Australia

S. K. Paul et al. (eds.), *Supply Chain Risk and Disruption Management*, Flexible Systems Management, https://doi.org/10.1007/978-981-99-2629-9_9

Australian agribusiness is export-oriented and operates on a large scale. However, its considerable distances linking production, markets, inputs, and infrastructure make it a high-cost operator relative to international competitors: Australian grains logistic costs are reckoned to be the highest in the World.

A succession of shocks to both demand and supply over the last decade has challenged the Australian agri-food supply chains (SCs) in the forms of natural disasters, geopolitical maneuvring in trade policies, a pandemic, and drought. These shocks are in many cases associated with broader trends such as climate change, expanded biosecurity threats, more fickle consumer needs, and changes in labor mobility. Australia's highly concentrated food retail and farm input industries also suggest that these shocks may be transmitted along the supply chain in complex ways.

In common with most industries, food supply chain management has adhered to conventional performance metrics associated with cost, logistics, and capacity utilization. Digital technology has rapidly matured and is providing a plethora of entirely new design options for the supply chain. The practical applications of these technologies feature improved productivity, targeting quality and quantity to the market while retaining logistic efficiency. The shocks outlined above call for unique ways to accommodate the Australian commercial and physical environment and farm and food management systems (Taghikhah et al. 2022). This commentary, however, refers to conditions not disturbed by the shocks outlined above, and the availability of benefits is heavily reliant on efficient supply chains to deliver them. That is to say, resilience in supply chains is mostly related to efficiency, and imperfectly represents a trade-off between costs and features of resilience such as redundancy.

This chapter focuses on aligning the definitions of agri-food supply chain resilience with management needs and highlights the potential applications of technologies and information systems in advancing resiliency. It outlines a set of shocks or challenges associated with different commodity sectors and identifies thematic benefits of resilience at varying echelons of supply chains.

2 Resilience and Risk in the Agri-Food Supply Chain

2.1 Supply Chain Resilience: Definitions and Management Approaches

Resilience is the ability of a system to maintain key functions and processes in the face of stresses or pressures by resisting and then recovering or adapting to change. Relevant definitions applied to agri-food systems include the collective ability of agri-food SC stakeholders to ensure acceptable, sufficient, and stable food supplies, at the required times and locations, via accurate anticipation of disruptions and the use of strategies that delay impact, aid rapid recovery, and allow cumulative learning post-disruption.

Paraphrasing Stone and Rahimifard (2018), agri-food resilience is the collective ability of agri-food SC stakeholders to maintain function by anticipating disruptions and using strategies that delay impact, aid rapid recovery, facilitate adaptation, and allow cumulative learning. The adaptive capacity of a resilient system is key in our definition: we do not require the system to bounce back to its original state or condition because we do not assume that the system persists. Therefore, the way the system performs can change as the system evolves toward its new state.

Currently, the focus of SC resilience literature is largely on proposing strategies for the post-disruption period (Rahman et al. 2021). A research gap then appears regarding strategies and recovery plans that can increase the resilience of supply chains during disruptions. In terms of design, static supply chain network configurations are insufficient to build readiness and responsiveness to deal with the ongoing disruption in the extraordinary situation (Hobbs 2020; Paul and Chowdhury 2020). Rather, SC designs should be adaptive and dynamic to address the disruption and sustain society and economy (Baveja et al. 2020; Sharma et al. 2020).

Maintenance, sustenance, and continuity are aspects of resilience most often related to the sustainability of resources, systems, conditions, and relationships. Solow (1991) proposed that a system is sustainable as long as its total capital is equal or greater in every succeeding generation. Costanza and Daly (1992) argued that sustainability only occurs when there is no decline in natural capital. The definition from the Brundtland Commission, which defined sustainable development as that which meets the needs of the present generation without compromising the ability of future generations to meet their own needs (Brundtland and Khalid 1987), is vague as "needs" are not defined. Costanza and Daly's (1992) definition emphasized the system's properties:

> Sustainability implies the system's ability to maintain its structure (organization) and function (vigor) over time in the face of external stress (resilience).

Trust is a system property and a widely used term that has gained in importance due to the growing data intensity of the digital economy, a number of high-profile trust breaches, and an overall shift from trust in institutions toward trust in communities (Botsman 2017), such as the agri-food SC. However, a shared understanding of trust—particularly along supply chains—is still in its infancy. Mayer et al. (1995) provided one of the most widely used frames of reference for trust when they proposed breaking down trust into ability, benevolence, and integrity. An alternative and easily deployed view of trust defines as "confidence with uncertainty." Thus, designing and managing a trusted supply chain can be broken down into the management of uncertainty and confidence management.

The lower the uncertainty in a SC, the higher the trust will be. Uncertainty management subdivides uncertainty into four facets:

- Systemic uncertainty—The variation in performance describes the actual systematic uncertainty of a supply chain (e.g., how predictable is the delivery date?). Dealing with systemic uncertainty is a significant challenge in high frequency, well synchronized just-in-time supply chains.

- Behavioral uncertainty—The more people-intensive (manual) a supply chain is, the higher will be the risk of variation in the performance of the individuals involved. Common standards, scripts, well-defined business rules, compliance standards, checklists, etc., are approaches to address behavioral uncertainties.
- Perceived uncertainty—As trust is a subjective assessment, incomplete data and information asymmetries can lead to perceptions different from reality.
- Vulnerability—While uncertainty describes possible shortcomings of a supply chain, vulnerability is about what happens when the supply chain fails. The assumption is that trust in a supply chain increases if the vulnerability is managed well.

Data-driven systems offer the opportunity to increase the volume of information available in the agri-food SC and engender trust by way of transparency. Profitable use of this information and associated technologies requires awareness along the SC of its value and function, and confidence in its use to demonstrate trust. Confidence boosters include:

- Democratic trust—Confidence can be built via mass data rating systems in which SC customers have a chance to assess the performance or via actual, shared measures (e.g., average on-time delivery).
- Special trust—This type of trust boosting is relevant for customers with unique requirements.
- Local trust—Local trust relies on feedback from trusted partners, i.e., a supply chain customer might trust another known customer.
- Global trust—Endorsement by third parties can raise trust in a SC. This is a common approach in supply chain maturity assessments.
- Institutional trust—If the provider of the supply chain is trusted, this trust might adhere to the supply chain itself. This type of trust-building is often long-term and more about brand reputation, etc., than informational approaches.

Approaches for increasing resilience include the Taguchi method (Mo and Harrison 2005), which minimizes variance in the outputs (at the operational and tactical level), and the risk management method (Gaonkar and Viswanadham 2007) which minimizes variability across the SC, especially in suppliers (at the operational, tactical and strategic level). One of the fundamental conflicts occurs between supply resilience on one hand and efficiency on the other (Christopher and Peck 2004): SC management has for a generation targeted reduced redundancy and shortened lag times. The systems design approach of Falasca et al. (2008) targets "efficient resilience," or resilience achieved at the lowest costs.

In terms of collaboration, collecting, sharing, and integrating data along the value chain are initial steps. Emerging data mining and process mining methodologies can be adapted to provide decision support to managers. For example, automatically monitoring and analyzing SC performance data by using a multi-perspective analytical approach can enhance performance (Keates et al. 2020). The adoption of real-time data sharing and data analytics across different organizations can assist with building trusted and resilient SCs. Various technologies such as the Internet of Things, Artificial Intelligence (AI)/Machine Learning algorithms, Blockchains, Cloud-based Business Intelligence, etc., can make data-driven decision-making a reality in near real time.

2.2 *Supply Chain Risk: Definitions and Management Approaches*

SC risks are generally classified into two major categories—micro and macro risks (Brusset and Teller 2017). Micro risks include operational risks caused by day-to-day uncertainties in standard operational activities. Risks due to large-scale disruptions and incidents such as natural calamities and economic crises are referred to as macro risks (Baghersad and Zobel 2021). A recently added third category, extraordinary risks, has been associated with pandemics impacting operations on a global scale.

Mathematical modeling (mainly optimization) and statistical modeling (mostly structural equation modeling) methods are the most common approaches to examining the impact of disruptions on supply chain performance. There have been limited applications of simulation modeling techniques to assess the extent of extraordinary disruption and evaluate the effectiveness of mitigation strategies.

Management issues center on observed interdependencies between organizations and their supply chains (Christopher and Peck 2004). Notably, the business may be at risk from its supply chain; or the SC from a business. Table 1 lists recovery strategies associated with shocks manifesting at different points in the supply chain.

3 Agro-Food Supply Chains' Vulnerabilities to Shocks

Key features of the six Australian food and farm sectors are summarized in Table 2. We discuss each in turn with regard to their supply chain vulnerabilities.

3.1 *Grains*

Grains and pulses constitute Australia's second-largest rural industry in terms of production value and export value, and are produced in a variety of landscapes located far from processing capacity and domestic consumption locations (Australian Bureau of Agricultural and Resources Economics and Sciences 2020; GRDC 2020). Concentration is apparent at downstream stages of the supply chain, including ownership of infrastructure. Well-developed trading operations with strong links to world markets occur all the way along the supply chain, predominantly involving traders who in turn own supply chain infrastructure (Stretch et al. 2014). The main grain handlers

Table 1 Supply chain risk and recovery strategies

Disruptions at levels of supply chains	Potential recovery strategies
Demand disruptions	Predicting and understanding buying tendencies by simulating consumer behavior
	Using social media to predict and guide demand
	Running models of consumer behavior to design the most effective strategies of market and non-market mechanisms to inform and guide demand
	Predicting demand algorithm specific to the product
	Wise labeling and awareness campaigns
	Confirming short-term demand-supply synchronization
Manufacturing disruptions	Flexible working arrangement
	Information and resource sharing among manufacturers
	Diversification of manufacturing plants in different locations
	Establishment of an emergency operation center for essential items
Supply disruption	Collective emergency sourcing
	Multiple backup suppliers
	Designing functional substitutes with similar nutritional values
	Accounting for natural capital and long-term trends in environmental quality (soil, biodiversity, water, etc.) that impact agricultural production
	Increasing redundancy by decreasing demand through promoting healthy lifestyles and ecological awareness, reducing waste and inefficiencies
Logistics disruption information and financial disruption	Alternative outbound and inbound logistics options
	Planning for local substitutes
	Collaborative transportation management and backup depot facilities
	Blockchain technology integration

are Cooperative Bulk Handling (CBH) in WA, GrainCorp in the Northern region, and Viterra, Emerald, and GrainCorp in the Southern region (Li et al. 2019).

The industrial organization of the grains sector sees highly concentrated grain export businesses, which also own significant proportions of the infrastructure for storage, freight, and export port operations. Much of this change is due to recent industry deregulation (Kalisch Gordon et al. 2016). Grain trading operations occur all along the supply chain so that grain in transit is predominantly owned by traders who also own the infrastructure. Supply chains in WA and SA are primarily structured to deliver grains to ports for export. In contrast, around 50% of the grain produced in eastern Australia is consumed locally, being processed into a variety of products.

Table 2 A summary of key features of the six Australian food commodity groups (inspired by KPMG (2020)).

	Grains	Horticulture	Dairy	Red meat	Seafood	Wine
Farm gate value ($bn)	12.6	10.0	4.4	65.7	5.3	45 (including wine)
Locations	Widespread, diverse physical and climatic conditions	Widespread, concentrated in small areas in most states	Concentrated in Southeast	Widespread, on less fertile land	Widespread, a variety of production formats	Numerous locations, specific soil, and climatic conditions
Market orientation	Major exporter significant domestic market in the eastern states	30% of production is processed. Major competition from imports	Limited exports	Major exporter	Wide variety of fresh, processed, and frozen products major exporter	Major exporter several market segments
Industrial organization	Highly concentrated downstream. environment and trading infrastructure	Faces highly concentrated retail environment	Rapidly consolidating, strong co-operative sector. Faces highly concentrated retail environment Excess processing capacity	Large foreign firms dominate processing capacity Domestic market dominated by concentrated retail	Independent retail co-exists with supermarkets Strong import competition	Mix of large and small at production and processing levels A few large companies with many established brands
Supply chain configuration	Dominated by road and rail from remote production locations Increased on-farm storage	Dominated by road and rail Dispersed processing and packaging facilities	Supply-demand imbalance requires long-distance transport outward from Victoria Reliance on cold chain	Rapid development of (cattle) feedlots	Complex supply chain including air transport. Reliant on cold chain	Demanding logistics at harvest Seasonal/vintage storage and distribution

(continued)

Table 2 (continued)

	Grains	Horticulture	Dairy	Red meat	Seafood	Wine
Risks and resilience associated with						
One health	Biosecurity • Well-developed quarantine system • Clean-green image maintained by traceability and provenance systems Reliance on agrichemicals; Vulnerability to labor, input, and product movement controls	Biosecurity • Well developed Food safety • Product integrity threats • Strong image maintained by traceability and provenance systems Long market recovery periods for fruits; Extreme dependence on seasonal labor	Food safety • Well developed, but vulnerable at several points Animal welfare • Strongly related/ perceived surrounding animal health and husbandry Environmental impacts • Strong perceptions regarding GHG emissions, water use, and waste management	Animal welfare • Strongly related/ perceived surrounding animal health and husbandry Environmental impacts • Strong perceptions regarding GHG emissions, water use, and waste management	Numerous Biosecurity threats to farmed and wild catch fish	Biosecurity • Well developed Food safety • Product integrity threats • Strong image maintained by traceability and provenance systems • Product integrity Long recovery periods for vineyards
Drought	Changed pest and disease burden Soil and water resource vulnerability Altered domestic trade patterns	Changed pest and disease burden Dependence on insect pollination Widespread reliance on irrigation water	Major vulnerability to feed availability and high costs Feed diversification to pasture exposes farms to further drought risk	Productivity, environmental performance, pasture resilience, and disease status are all vulnerable	Drought affects water quality for onshore fisheries and ocean habitats Fish life cycles are affected	Vulnerable to rainfall and related conditions Many locations reliant on irrigation

(continued)

Table 2 (continued)

	Grains	Horticulture	Dairy	Red meat	Seafood	Wine
Natural disasters	Vulnerable transport and handling infrastructure Exposure due to seasonal production patterns. Export markets' supply chains are vulnerable	Vulnerable transport and handling infrastructure Exposure due to short harvest period. Export markets' supply chains are vulnerable	Vulnerable transport and handling infrastructure, particularly farm collection points	Vulnerable transport and handling infrastructure, particularly farm collection points. Export markets' supply chains are vulnerable	Vulnerable transport and handling infrastructure, particularly farm collection points. Export markets' supply chains are vulnerable	Vulnerable to smoke damage from bushfire Logistics infrastructure vulnerable to floods and bushfire
Geopolitics	Vulnerable to capricious trade regulation High supply chain costs	Heavy reliance on the Chinese export market Vulnerable to visa and mobility rules on labor	Volatile international markets. Non-transparent pricing on domestic markets	Heavy reliance on export markets, with access subject to NTB Likely linkage of trade to environmental performance	Strong competition from less regulated competitors. Resource access subject to threats. Poor recognition of provenance for most products	Vulnerable to capricious trade regulation High supply chain costs
Technology	Dispersed storage and far-flung transport routes challenge quality and stewardship	Data streams vulnerable to cyber-attack Limited skills in technology use	Data streams vulnerable to cyber-attack Underdeveloped pricing system raises uncertainty	Technology adoption is slow across the sector	Low uptake across the sector. Vulnerable to cyber-attack	Well-developed uptake but data streams vulnerable to cyber-attack

The Australian grains industry is vulnerable to biosecurity incursions or food safety breaches in several senses including delays in the importation of inputs (chemicals and fertilizers), reduced quantity/quality of grain produced, rejection of deliveries to bulk handlers, higher storage, and handling costs, and even rejection of deliveries to export markets. Increased environmental accountability is a concern for the industry, particularly its heavy reliance on glyphosate and the potential consequences for Australia's reputation as a clean and safe producer of grain. Traceability (e.g., food safety) and sustainability requirements (e.g., management of pests and disease/chemical use) create challenges in meeting consumer preferences while maintaining competitiveness (KPMG 2020). COVID-19 has heightened these concerns and also produced a shift in consumer demand away from food service toward eating at home, making the sector vulnerable to travel restrictions and quarantine, and their impacts on labor (ABC News 2020).

Increasing climate variability affects production and quality, and risks of pests and diseases, threaten the sector's performance and national food security (Kingwell 2019a). White et al. (2018) predict an associated production shift as lower-yielding locations further from the port are likely to become increasingly expensive relative to high-yielding locations nearer to the port. Drought is also associated with rural communities' viability and so creates vulnerability for supply chain participants' access to labor (KPMG 2020). Severe weather events threaten production, transport infrastructure, and digital connectivity (e.g., towers, sensors), with consequences exacerbated at certain times of the year (e.g., harvest). Storage and handling costs are scale-related and so susceptible to such shocks (KPMG 2020; Deloitte Access Economics 2019).

Trade barriers due to actual or perceived contamination or use of selected chemicals (e.g., reliance on Glyphosate; market access for GM) or tariffs (e.g., barley tariffs and China, chickpea tariffs and India) can potentially lead to loss of market access or reduced international competitiveness (KPMG 2020). Australian supply chain costs are higher than most of its competitors which impacts international competitiveness. Australia's grain export competitors are undertaking major investments in their supply chains and combined with attractive sea freight rates pose market access threats (White et al. 2018). Low-cost competitors South America (i.e., Brazil and Argentina) and the Black Sea region (Russia, Ukraine, and Kazakhstan) have significantly increased their grain production and grain exports in recent years (GRDC 2020).

Recent growth in on-farm storage has changed supply chain cost structures and risk allocation along the supply chain (Kingwell 2019b). With on-farm storage, the risks associated with grain hygiene, grain classification, and grain ownership remain with the grower rather than being transferred to grain handlers. Stewardship obligations regarding grain quality become increasingly important to preserve Australia's reputation as a reliable supplier of safe, high-quality grain (White et al. 2018).

3.2 Horticulture

Australian horticulture is dominated by fruit and vegetables (over 70% of value) as well as amenity production (including turf, nursery, and cut flowers). Concentrated in limited areas of most states, the sector is a major employer (Innovation 2020) both for harvest and crop husbandry and for downstream processing and value addition. Maximum shelf life, pest and disease management, and protection of perishable products are key for the products' quality, safety, pricing, and market access—particularly in markets supplied as out-of-season imports. Biosecurity incursions or health pandemics can particularly create disruptions to air freight and in-market cold chain logistics and to the availability of farm inputs (e.g., chemicals). Product sabotage (recently experienced for packages of strawberries) and other contaminations at the farm, packing or retail stages are industry points of vulnerability. Although heavily regulated, KPMG (2020) notes that food safety and audit processes are typically paper-based and time-consuming.

Travel restrictions and quarantine periods associated with the Covid-19 outbreak highlight the extent of potential industry disruption due to health events. Australian horticulture is heavily dependent on access to international labor, and widely reported large shortfalls in labor shortages expose the industry to risks of wasted products and disruptions to food supply (ABC News 2020).

Changes in climate are increasing the risk to agricultural outputs by promoting conditions that favor pests and diseases. If these threats infiltrate regions designated as Pest Free Areas (PFAs), their status could be compromised, which in turn would limit their access to global markets. Consequences include crop destruction, incursion management costs, and market access restrictions/bans. The Varroa mite parasite is a big biosecurity risk for the industry as it can decimate bee colonies which contribute some $1.6 billion annually as pollinators (Innovation 2020). Bees are also directly susceptible to droughts and floods, and water availability. Severe weather events threaten production sites, and value addition and transport infrastructure, with long delays between assets' destruction and their re-establishment to full production.

Travel and visa problems disproportionately impact horticulture's labor supply, particularly at key times of the year (ABC News 2020). Tariffs and non-trade barriers impact competitiveness and export market access, and Australian horticulture's specific exposure to the Chinese market has highlighted these concerns due to disruptions to other sectors serving that market. KPMG (2020) identify several vulnerabilities related to technology, including hacking of supply chain data and data governance issues limiting data sharing along the supply chain. In addition, industry skills in technology use in the sector have been recognized.

3.3 Dairy

Dairy production takes place in concentrated locations in all Australian states but is predominantly located in the southeast corner of the country. Many farms are in coastal areas and rely on rainfall for pasture growth, while others are dependent on irrigation schemes. The number of dairy farms in Australia has been gradually declining with a shift toward more intensive operating systems with larger economies of scale (Dairy Australia 2019) and associated reliance on specialized, and perhaps vulnerable, feeding systems.

The industry is reliant on value addition through preservation and processing, and high levels of supply chain interdependence and trust due to milk's continuous flow. This has resulted in cooperative ownership at some locations. International acquisitions have also recently been a feature of the processing stage, particularly as a source of capital for investment in new equipment (Dairy Australia 2019). Food safety and pest and disease management—both crucial for market access—are particular vulnerability and has further narrowed the focus of transaction relations between producers and processors, and the highly concentrated retail sector. COVID-19's constraining effects on labor availability have strained these chain relationships.

Public sentiment imposes on dairy in terms of animal welfare, as well as environmental and human health impacts. These relate to industry practices (bobby calves' slaughter, animal handling, and treatment) as well as GHG emissions (Dairy South Australia 2017). Other environmental concerns include waste management and land and water degradation.

Drought in dairy areas and beyond imposes risks both to animal production, and to feed and water costs and supplies (Dairy Australia 2019). Longer-term climate variation increases the vulnerability of pasture systems, and the affordability of water as competing uses (including that from other food and farm sectors) appear. Severe weather events have the potential to damage rail, road, and storage infrastructure (KPMG 2020), particularly in terms of cold chain logistics disruptions and the availability of farm inputs.

Volatility in farmgate milk prices and farm incomes in recent years have constrained industry growth (Dairy Australia 2019). Related scheduling of products into markets both domestic and international, has been problematic in terms of imbalances of fluid milk demand between states and in response to price volatility. A lack of transparency in this market means that pricing signals are not disseminated back to milk producers. Many manufacturers have the excess capacity as plants are underutilized. Dairy is susceptible to cyber threats by way of disruptions to supply chain flows and product quality and safety assurance. Exports may be threatened if satisfactory data guaranteeing the food safety of products is not available (KPMG 2020).

3.4 Red Meat

Red meat is Australia's single largest rural industry and employs around 400 thousand people (ABARES 2020; Meat and Livestock Australia 2020). Comprising several subsectors (lamb and mutton, goats, beef, veal, pigs), and spread over large areas of the country (Meat and Livestock Australia 2019), long delivery distances for sales and processing provide logistic vulnerability. Biosecurity is a serious concern for the domestic industry and for access to international markets. Contact pathways, and later labor shortages, associated with COVID-19 have visited significant disruptions in the meat supply chains at several levels. Changes in consumption of meat due to health concerns have seen shifts between trends in location, volume, and form of purchases, particularly in light of emerging sources of protein (KPMG 2020).

Confronting the red meat industry are public sentiment concerns over animal welfare and environmental and human health impacts. Key animal welfare issues to relate animal handling and husbandry practices which can decrease productivity and damage social license. As with dairy, red meat contributes a significant proportion of Australia's agricultural GHG emissions. The red meat and livestock industry contributes 10% of Australia's GHG emissions and is committed to carbon neutrality (CN30) by 2030 (Meat and Livestock Australia 2020). Other concerns center on the use of antibiotics and hormonal growth supplements.

Prolonged dry periods can have short- and long-term impacts on animal productivity, and also on soil quality/health and environmental performance (topsoil erosion) that impacts red meat production. Further, the facility for management of grazing systems during droughts has been identified as a vulnerability (KPMG 2020). Drought's impacts on rural communities also restrict labor supply to the industry, and these are exacerbated by severe weather events such as floods which damage transport and processing infrastructure, and access to markets (KPMG 2020). Natural disasters can also result in unintended animal welfare issues whereby livestock are killed directly or left injured and maimed. More than $800 m worth of livestock was lost in the North Queensland floods in early 2019.

Australia's reliance on certain export markets increases its susceptibility to geopolitical shocks. Examples include the (animal welfare-based) live cattle export ban to Indonesia in 2011, and non-compliance-related rejections of processed meat by China in 2019. Australian Meat Processor Corporation (2020) identifies market access as a key industry consideration, also noting the EU's import quotas. NTBs impose additional costs on the red meat value chain and most notably have pronounced impacts in China, the Middle East, and Southeast Asia. Regulatory and compliance costs across the red meat supply chain are already seen as high and their further escalation is a threat. The utilization of information in general, and digital tools in particular, is low. This particularly applies to information flow back up the supply chain to producers with regard to product attributes and compliance issues.

3.5 Seafood

The Australian seafood sector employs around 41 thousand people and is spread across diverse climatic regions and ecosystems including ocean, estuary, and river fishing. Tasmania, Western Australia, and South Australia combined represented 68 percent of gross production value in 2018. The industry has a complex structure with participants ranging from small family-owned businesses to large corporate operations. Australia predominantly exports high-value products and imports low-cost chilled and frozen products. In contrast to other sectors, there are many independent fishmongers operating in the retail sector in competition with major grocery chains. The distribution channels for fresh seafood are convoluted due to the diverse range of species, perishability, range of products, expanse of geographic supply and production sources, various requirements for early-stage cold chain handling, and differing market outlets (Spencer 2016).

Biosecurity incursions of particular concern for the seafood industry are diseases in fish stocks affecting supply. Disease in wild catch is hard to monitor and control and has wide-scale implications for both fish health and food safety. Health events are likely to create fishing, processing, and logistics node interruptions with potential disruption/shutdown of points of distribution and point of sales. Logistical/supply chain issues include the closure of freight and logistic routes, limited cold chain availability, disruptions to airfreight, and labor shortages (Seafood Industry Australia 2020).

Drought exerts substantial stresses on land-based and ocean fisheries, as well as perceptions of industry sustainability and social acceptance (Seafood Industry Australia 2020). Drought impacts the ability to harvest, particularly in wild catch fishing zones. Longer-term climatic change impacts fish species and breeding, particularly in species with breeding cycles related to rivers. Transport and processing infrastructure—which is frequently coastal—is susceptible to natural disasters (e.g., flood and bushfires) with potential disruptions to supply chains and logistics, particularly cold chain storage and availability/capacity. Many live and fresh fish products require cold chain capabilities, which are in limited supply across Australia. Implications arise for supply chain integrity and quality maintenance. The seafood industry also has a heavy reliance on certain export markets that have limited cold chain infrastructure. Climatic events can impact fish species and breeding and harvest capabilities, particularly in wild catch fishing zones. Protection of resource access and property rights is paramount to the confidence and growth of the Australian seafood industry. Any erosion of secure access to resources, both aquatic and terrestrial, is a major impediment (Seafood Industry Australia 2020).

Australian fish faces strong import competition from low-cost production in Asia and has a heavy reliance on certain export markets for industry returns and growth. Market access closures can then significantly impact both production and processing (KPMG 2020). Access to labor is problematic and related to migration and visas. Food fraud is widely recognized as a significant problem, and technological advance has increasingly addressed provenance: constraints have included limited trust in data systems along the supply chain.

3.6 Wine

Australia's 6,000 vineyards and 2,500 wineries support around 164,000 jobs and export around 60% of production (KPMG 2020). The industry is present in limited areas of most states, featuring a range of grape types. Its key challenges include biosecurity, exacerbated by the traffic associated with growth in wine tourism, and chemical resistance to pests and diseases. These considerations are, however, central to market access and consumer enthusiasm for the product (KPMG 2020). Periodic swings in consumer preferences impose vulnerability on producers who must make long-term decisions on varieties grown. Provenance is a traditional attribute of wine systems, which is being reinforced by digital technologies. This is in turn vulnerable in terms of cyber security concerns and mistrust between supply chain actors.

Drought is a threat to production quality and quantity, and broader climate variability affects water availability and temperature to the extent of forcing shifts in vineyard locations. The industry's location also predisposes to competing uses of land and water. Extreme weather events can disrupt production, transport and logistics infrastructure, employment, and market access with impacts on international competitiveness and perceptions of trust and reliability in Australia as a "clean, green and sustainable" producer of wine-environmental stewardship. The 2020 bushfires in eastern Australia are estimated to have cost the industry around $100 million by these mechanisms.

Exposure to international markets features reliance on the Chinese market, to which access was severely curtailed in 2020 and 2021. The world market is subject to oversupply, with ramifications for price, supply, and inventory, and interruptions to trade flows and market access due to trade barriers (KPMG 2020). Travel restrictions and quarantine periods/changes to visa requirements have impacted labor supply which particularly impacts harvest. Data management and reporting, traceability, counterfeiting, data breaches, and fraud are all issues that potentially compromise perceptions of trust and reliability. Lack of internet, communications, and weather-monitoring infrastructure are all identified as challenges for the industry.

4 Digital Technologies and Their Contribution to Resiliency in the Agro-Food Supply Chain

Agri-food supply chains feature uncertainties and complexities that predispose them toward vulnerabilities not seen in other sectors. Challenges to managers then emerge, related to freshness and food safety, shelf life, specific considerations for food logistics and handling, and processing times. Issues related to production seasonality, variability of supply (farm-based operations) quality and quantity as well as unpredictable consumer demands require a holistic approach (both technology and management decisions) and consistent efforts in the long term. Digital technologies have offered substantial opportunities concerned with data flow along and between chains, and information transparency for the delivery of credence products. To enhance resilience, these technologies need to be better utilized for their shortened management response times (e.g., for biosecurity threats to the grape crop), high-level market co-ordination (e.g., for scheduling milk flows to processing capacity), and heightened provenance (e.g., freshness and origin of fish products).

Adoption of advances in predictive and prescriptive analytics capabilities in agri-food supply chains requires that they be decision-ready and demonstrate a value proposition (Griffith et al. 2013; Lezoche et al. 2020; Rejeb et al. 2021). Big data and supply chain performance assessment have seen applications of artificial intelligence (Taghikhah et al. 2022) and machine learning techniques (Taghikhah et al. 2021) to analyze consumer preferences and behavior (Taghikhah et al. 2020). Agent-based modeling has been used to understand consumer behavior (Taghikhah et al. 2021) and compare policies and market and non-market mechanisms (Taghikhah et al. 2022). The adoption of blockchain technologies in agri-food supply chains can foster collaboration, increase transparency, and build trust. It also influences identified resilience strategies, such as advanced labeling and product identification, the use of scannable QR codes, and information about ecological and carbon footprints associated with products, and chips and biomarkers to monitor food condition and freshness(Rahman et al. 2022). Social media offers additional ready-to-use platforms and tools for demand management, including product promotions, exploitation of shifts in lifestyle and health concerns, and food waste. Decision support tools are associated with predictive analysis, optimisation, and risk analysis, as well as visualization.

Collaboration along the agri-food supply chain is central to realization of benefits from many data-driven solutions. Trust plays an important role in collaboration not only among supply chain actors using data, but also in terms of ownership and control of data. The integration of farm management decision-making tools with supply chain analytics has been limited(Taghikhah et al. 2019); there is a missing feedback loop between market demand, prices, and farm management decisions, what and how much to produce, and what practices to adopt (Taghikhah et al. 2021). Modeling, especially participatory modeling, is essential to understand the various scenarios and management solutions. Miranda-Ackerman and Azzaro-Pantel (2017) and Jonkman et al. (2019) offer recent examples of models addressing a range of decisions from farm level (e.g., organic versus conventional farming) to the production (e.g., technology selection) and distribution level (e.g., transportation route). Although studies addressing SC decisions simultaneously are still lacking, the literature trend is toward more integrative, holistic agri-food models.

Reviews of commodity sectors' status and vulnerabilities generate a set of potential research gaps for advances associated with digital solutions. Not all vulnerabilities are suitable for data-driven solutions, and uptake of existing digital technologies is apparently afforded priority over the development of new technologies. Adoption is in turn constrained by data governance, skills, and the lack of a compelling value proposition, although this does not easily constitute a researchable problem.

5 Conclusion

Factors contributing to supply chain resilience have been variously identified and defined in the literature, although a unified empirical approach is lacking. Supply chain performance in general has increasingly and successfully targeted costs and efficiency and has been implemented alongside the exercise of market power. These two features of supply chain management have respectively prevented the development of two generic approaches to supply chain resilience: redundancy and collaboration along the supply chain. Beyond these generic terms, a small research literature has developed around supply chain resilience, and it has entered mainstream supply chain management discourse as a part of the SCOR supply chain management initiative.

Data-driven approaches to the analysis of supply chain resilience center on dynamics and feedback. Management interpretation of this research lists actions and indicators which contribute to different aspects of resilience. A large number of such indicators, and associated metrics, are available.

The current chapter focused on a number of agri-food industry vulnerabilities to macro risks: both general across industries and specific to some individual industries or supply chain stages. It also reflected on the potential for data-driven solutions to enhance resilience and the prioritization of knowledge gaps that constrain improvements in resilience.

With regard to the trends and applications of resistance analytics in supply chain modeling (Golan et al. 2020), several recommendations are listed as:

- (Re)consideration of the definition of supply chain resilience across all supply chain models and sectors is necessary to make resilience management more efficient.
- Consideration of different types of disruptions within supply chain resilience models—especially assessing system recovery from unknown disruptions and systemic threats—is necessary to expand the scope that supply chain resilience management is able to quantify.
- Consideration of the tiered approach to modeling, ranging from simple metrics to advanced network models, is necessary for understanding which quantification method to apply to the analytic need.
- Consideration of the supply chain within the broader context of other networks that constitute value generation (e.g., command and control, cyber, transportation) is necessary for the quantification of global network interactions and more

robust supply chain resilience models that accurately portray trade-offs between efficiency and resilience to avoid cascading failures.

References

ABARES (2020) Agricultural Commodities: June 2020. Australian Bureau of Agricultural and Resources Economics and Sciences. https://doi.org/10.25814/5ec210973b2b8

ABC News (2020) Coronavirus restrictions put pressure on upcoming bumper northern NSW grain harvest. https://doi.org/ https://www.abc.net.au/news/2020-09-22/north-west-farmers-prepare-for-harvest/12685056

Australian Bureau of Agricultural and Resources Economics and Sciences (2020) Agricultural Commodities

Australian Meat Processor Corporation (2020) Strategic plan 2020–2025. https://www.ampc.com.au/uploads/2020-25-Strategic-Plan.pdf

Baghersad M, Zobel CW (2021) Assessing the extended impacts of supply chain disruptions on firms: an empirical study. Int J Prod Econ 231:107862

Baveja A, Kapoor A, Melamed B (2020) Stopping Covid-19: a pandemic-management service value chain approach. Ann Oper Res 289(2):173–184

Botsman R (2017) Who can you trust?: how technology brought us together—And why it could drive us apart. Penguin UK

Brundtland GH, Khalid M (1987) Our common future. Oxford University Press, Oxford, GB

Brusset X, Teller C (2017) Supply chain capabilities, risks, and resilience. Int J Prod Econ 184:59–68

Christopher M, Peck H (2004) Building the resilient supply chain

Costanza R, Daly HE (1992) Natural capital and sustainable development. Conserv Biol 6(1):37–46

Dairy Australia (2019) In focus 2019: the Australian dairy industry. D. Australia. https://www.dairyaustralia.com.au/resource-repository/2020/07/09/australian-dairy-industry-in-focus-2019#.X4UEAmgzaUk

Dairy South Australia (2017) Understanding greenhouse gas emissions on your farm. https://www.naturalresources.sa.gov.au/files/sharedassets/adelaide_and_mt_lofty_ranges/land/greenhouse-gasesdairy

Deloitte Access Economics (2019) The impact of freight costs on australian farms. A. Australia

Falasca M, Zobel CW, Cook D (2008) A decision support framework to assess supply chain resilience. In: Proceedings of the 5th international ISCRAM conference

Gaonkar RS, Viswanadham N (2007) Analytical framework for the management of risk in supply chains. IEEE Trans Autom Sci Eng 4(2):265–273

Golan MS, Jernegan LH, Linkov I (2020) Trends and applications of resilience analytics in supply chain modeling: systematic literature review in the context of the COVID-19 pandemic. Environ Syst Decis 40(2):222–243

GRDC (2020) Australia's grains industry in 2030—A look into the future

Griffith C, Heydon G, Lamb D, Lefort L, Taylor K, Trotter M, Wark T (2013) Smart farming: leveraging the impact of broadband and the digital economy. CSIRO and University of New England, New England

Hobbs JE (2020) Food supply chains during the COVID-19 pandemic. Can J Agricul Econ/Revue Can D'agroecon 68(2):171–176

Innovation H (2020) Horticulture statistics handbook 2018/19. In

Jonkman J, Barbosa-Póvoa AP, Bloemhof JM (2019) Integrating harvesting decisions in the design of agro-food supply chains. Euro J Oper Res 276(1):247–258. https://doi.org/10.1016/j.ejor.2018.12.024

Kalisch Gordon C, McKeon D, Whiteley C, Southan M, Price C, MacAulay G (2016) State of the southern region grains industry and challenges to innovation & growth. S. a. C. Grain Growers Ltd.
Keates O, Wynn MT, Bandara W (2020) A Multi Perspective framework for enhanced supply chain analytics. In: International conference on business process management
Kingwell R (2019a) Australia's Grains Outlook 2030. A. E. G. I. Centre
Kingwell R (2019b) International grain supply chains: issues and implications for Australia
KPMG (2020) Mission food for life. F. Agility
Lezoche M, Hernandez JE, Díaz Md.MEA, Panetto H, Kacprzyk J (2020) Agri-food 4.0: a survey of the supply chains and technologies for the future agriculture. Comput Ind 117:103187
Li K, Kingwell R, Griffith G, Malcolm B (2019) Measuring the returns to investment in RD&E; in the WA grains industry using equilibrium displacement modelling. Aust Agribus Rev 27(1673-2019-3373):65–106
Mayer RC, Davis JH, Schoorman FD (1995) An integrative model of organizational trust. Acad Manag Rev 20(3):709–734
Meat and Livestock Australia (2019) 2019 State of the Industry Report. https://www.mla.com.au/globalassets/mla-corporate/prices--markets/documents/trends--analysis/soti-report/mla-state-of-industry-report-2019.pdf
Meat and Livestock Australia (2020) 2020 State of the Industry Report. https://www.mla.com.au/globalassets/mla-corporate/prices--markets/documents/trends--analysis/soti-report/mla-state-of-industry-report-2020.pdf
Miranda-Ackerman MA, Azzaro-Pantel C (2017) Extending the scope of eco-labelling in the food industry to drive change beyond sustainable agriculture practices. J Environ Manag 204:814–824. https://doi.org/10.1016/j.jenvman.2017.05.027
Mo Y, Harrison TP (2005) A conceptual framework for robust supply chain design under demand uncertainty. In Supply chain optimization. Springer, pp 243–263
Paul SK, Chowdhury P (2020) A production recovery plan in manufacturing supply chains for a high-demand item during COVID-19. Int J Phys Distrib Logist Manag
Rahman T, Taghikhah F, Paul SK, Shukla N, Agarwal R (2021) An agent-based model for supply chain recovery in the wake of the COVID-19 pandemic. Comput Ind Eng 158:107401
Rahman T, Paul SK, Shukla N, Agarwal R, Taghikhah F (2022) Supply chain resilience initiatives and strategies: A systematic review. Compu Indus Eng 108317
Rejeb A, Rejeb K, Zailani S (2021) Big data for sustainable agri-food supply chains: a review and future research perspectives. J Data Inf Manag 3(3):167–182
Seafood Industry Australia (2020) Strategic Plan 2018–2023. https://seafoodindustryaustralia.com.au/about-us/strategic-plan/
Sharma A, Rangarajan D, Paesbrugghe B (2020) Increasing resilience by creating an adaptive salesforce. Ind Mark Manag 88:238–246
Solow RM (1991) Sustainability: an economist's perspective. In: Marine policy center woods hole, Massachusetts
Spencer S (2016) From farm to retail—How food prices are determined in Australia. In: Canberra: Australian Government Rural Industries Research and Development ….
Stone J, Rahimifard S (2018) Resilience in agri-food supply chains: a critical analysis of the literature and synthesis of a novel framework. Supply Chain Manag: Int J
Stretch T, Carter C, Kingwell R (2014) The cost of Australia's bulk grain export supply chains: an information paper. Australian Export Grains Innovation Centre, Perth
Taghikhah F, Borevitz J, Costanza R, Voinov A (2022) DAESim: a dynamic agro-ecosystem simulation model for natural capital assessment. Ecol Model 468:109930
Taghikhah F, Erfani E, Bakhshayeshi I, Tayari S, Karatopouzis A, Hanna B (2022) Artificial intelligence and sustainability: solutions to social and environmental challenges. In: Artificial intelligence and data science in environmental sensing, pp 93–108. Academic Press
Taghikhah F, Filatova T, Voinov A (2021) Where does theory have it right? A comparison of theory-driven and empirical agent based models. J Artific Soc Soc Simul 24(2)

Taghikhah F, Voinov A, Filatova T, Polhill JG (2022) Machine-assisted agent-based modeling: opening the black box. J Comput Sci 64:101854

Taghikhah F, Voinov A, Shukla N (2019) Extending the supply chain to address sustainability. J Clean Produc 229:652–666

Taghikhah F, Voinov A, Shukla N, Filatova T (2020) Exploring consumer behavior and policy options in organic food adoption: insights from the Australian wine sector. Environ Sci Policy 109:116–124

Taghikhah F, Voinov A, Shukla N, Filatova T (2021) Shifts in consumer behavior towards organic products: Theory-driven data analytics. J Retail Consum Servic 61:102516

Taghikhah F, Voinov A, Shukla N, Filatova T, Anufriev M (2021) Integrated modeling of extended agro-food supply chains: a systems approach. Eur J Oper Res 288(3):852–868

White P, Carter C, Kingwell R (2018) Australia's grain supply chains: costs, risks and opportunities. South Perth: Australian Export Grains Innovation Centre

Prioritization of Risks in the Pharmaceutical Supply Chains: TOPSIS Approach

Rajesh Kumar Singh

1 Introduction

Risk is defined as "The effect of uncertainty on objectives" (International Labour Organization 2011). Every Organization or System is influenced by certain internal and external factors which induce uncertainties regarding the timely realization of its objectives (Lawrence et al. 2020a, b). This uncertainty's effect on an organization or a System's objectives is termed "Risk." This uncertainty of events leads to the occurrence of unfavorable outcomes like late deliveries, business losses, financial burdens, etc. (Mangla et al. 2016). It is observed that the impacts of the risk are personal to the concerned individuals. However, Risk in the Industrial Sector has an even greater impact since it hampers not only the industrial performance and returns but also the consumer-industry relationship. Moktadir et al. (2018) explained managing risks in the health service sector is a very important task and thus it needs high attention. Consequently, there has been an increase in the research nowadays to study risks in these industries mainly in the supply chain and to mitigate those risks. Over the past years, there has been a growing interest in Supply Chain Management (Manuj and Mentzer 2008) as it is seen as a key component of organizational effectiveness and competitiveness (Porter 1985; Womack and Jones 2005). In today's highly competitive world, organizations are more prone to vulnerabilities in their supply chain due to changing product demand, materials supply, skills, and equipment requirements (Finch 2004; Enyinda et al. 2009; Gandhi et al. 2016). Because of the above-stated reasons, managing risks holds top priority in any organization's priority list (Derrong Lin and Hertig 2021). It has been observed that large companies employ better risk management strategies (Ozdemir et al. 2022). Managing risks helps organizations deal with any disruption related to the supply chain (He et al. (2021); Christopher and Lee 2004; Tuncel and Alpan 2010; Mangla et al. 2016). Mouloudi and Evrard

R. K. Singh (✉)
Management Development Institute Gurgaon, Gurgaon, India
e-mail: rajesh.singh@mdi.ac.in

S. K. Paul et al. (eds.), *Supply Chain Risk and Disruption Management*, Flexible Systems Management, https://doi.org/10.1007/978-981-99-2629-9_10

Samuel (2022) discussed the economic and environmental circumstances involved in the accessibility and availability of raw materials which are the source of risks in supply chains.

Risk Management is developing as a major contributor to formulating decisions and exercising control in major fields of management (Lawrence et al. 2020a, b; Giannakis et al. 2004). Baryannis et al. (2019) highlighted that there is no specific or universal definition for supply chain risk management but emphasized that supply chain risk management should include identification, evaluation, mitigation, and monitoring of risks. Shenoi et al. (2018) encore supply chain risk management as an essential part of supply chains. Brindley (2004) mentions the importance of risk management in supply chain management in this modern era where the industries prefer to have a technological and competitive advantage over one another. Barroso et al. (2010) highlighted that resilient supply chains help organizations to be competitive. Singh and Singh (2019) explain that supply chains should be able to develop strategic responses to assess and mitigate the risks. El Baz and Ruel (2021) suggest cooperation among the partners involved in different stages of the Supply chain to find the sources of threats that affect the consequences of SCRM practices. Kilubi (2016) provides eight leading strategies to deal with supply chain resilience: redundancy, flexibility, postponement, visibility and transparency, multiple sourcing, joint planning and coordination, relationship/partnership, and collaboration. Kamalahmadi and Parast (2016) also laid importance on all the above strategies. Resilience Strategies and Recovery are the main mechanisms to build a resilient pharmaceutical supply chain (Derrong Lin and Hertig 2021; Blackhurst et al. 2011; Scholten et al. 2019). Daghfous et al. (2021) studied the importance of the risk management due to knowledge loss in Supply chains.

Elleuch et al. (2013) have combined both the quantitative and qualitative methods to identify, assess, and mitigate risks. Above these three processes (Sodhi et al. 2012) also consider responsiveness to the risk as a significant step. Furthermore (Elleuch et al. 2013) have also found five dimensions where improvements can be made in the pharmaceutical supply chains. These include budget consistency, reliability, fluidity of drug circuit, flexibility improvement, and bullwhip effect control to manage a large variety of items. To manage supply chain risks and to make them more resilient, the identification of supply chain characteristics is very important (Li et al. 2020a, b). These characteristics include Total Productive Maintenance (TPM) (Modgil and Sharma 2016) and Total Quality Management (TQM) (Sharma and Modgil 2020) in the Indian pharmaceutical industry. Sriyakul et al. (2019) examined the association of supply chain risk management strategies with SC integration and the performance of supply chains. Dubey et al. (2019) identified supply chain visibility, trust, supply chain connectivity, information sharing, and cooperation as the constructs to generate resilience in the supply chains. Tukamuhabwa et al. (2015) also laid importance on information sharing. Moreover, (Ivanov and Dolgui 2020; Dubey et al. 2020) added blockchain and supply chain 4.0 as the resilience generators.

Organizations should follow a formal structure to identify and quantify the risks (Khan and Burnes 2007; Mangla et al. 2016). Previous literature highlighted a joint decision-making strategy as a recovery mechanism but the latest works of literature focus on the coordination of decisions as a collaborative and proactive process for building resilient pharmaceutical supply chains (Derrong Lin and Hertig 2021; Scholten et al. 2019). Also, for smooth functioning in all the sectors of the global supply chains, businesses are required to enhance supply chain risk management strategies (Hohenstein 2022).

According to the Council of Supply Chain Management Professionals (CSCMP), a Supply Chain may be defined as "the material and informational interchanges in the logistical process stretching from acquisition of raw materials to delivery of finished products to the end-user." It has been observed that the Pharmaceutical Supply Chains, in particular, have been facing ever-increasing complexity of operations in the midst of strengthening regulatory and inflationary pressures. Supply chain management in the pharmaceutical sector has been challenging because its activities and products have direct interaction with the consumers' health (Lawrence et al. 2020a, b; Mustaffa and Porter 2009). Hence, the need for analyzing the risks in this sector gains equal importance.

Managing a Pharmaceutical Supply Chain is a complex task due to the dynamic network structure where the markets, processes, and products are closely involved (Jaberidoost et al. 2013). The pharmaceutical supply chain requires smooth processing of the flows as it deals with the medicines (also, medical components), and reaching medicines at the right time with the right amount and right quality is very important for any pharmaceutical supply chain (Lawrence et al. 2020a, b; Manuj and Mentzer 2008; Jaberidoost et al. 2015; Mehralian et al. 2012). Chopra and Meindl (2014) encore the importance of quality and regulatory compliance in a pharmaceutical supply chain. They emphasize information sharing to make the supply chain more effective and to fulfill customer requirements. The pharmaceutical supply chains are more prone to risks due to the economic, political, and social instability of the developing countries (Enyinda et al. 2009; Jaberidoost et al. 2015). Bigdeli et al. (2013) explained the importance of pharmaceutical supply chains due to the urgency, regulations, safety, and transportation of medicines. These supply chains hold importance because they can directly impact the healthcare system of any country (Hung et al. 2005; Tazin 2016). Hence, identifying risks affecting the pharmaceutical supply chains is very important. Identifying risks helps the organizations minimize the liabilities and costs, avoid waste, and enhance the efficiency of the supply chains (Goodarzian et al. (2020); Kwak and Dixon 2008; Rogachev 2008). It further helps the organizations to develop strategies to minimize such risks (Derrong Lin and Hertig 2021; Hulbert et al. 2008; Jaberidoost et al. 2013) and to increase the performance of the organization (Breen 2008; Mangla et al. 2015).

Several challenges are also associated with the pharmaceutical supply chains such as human resource dependence, sustainable supplier failure, short product lifecycle, warehouse management, inventory management, a quality problem with the product, lack of information, and seasonal demands for the products (Kuo et al. 2020; Craighead et al. 2007; Privett and Gonsalvez 2014; O'Connor et al. 2016; Luthra et al. 2017). However, Wahyu et al. (2022) showed that the objective of sustainable business can be achieved with the involvement of Human resources. In addition, technologies help organizations in mitigating the risks in the pharmaceutical supply chains (Mahendran et al. 2011).

The outbreak of the Novel Coronavirus in late December 2019 (Li et al. 2020a, b) caused the shortages of various commodities. The pandemic has caused major disruptions in domestic as well as global supply chain systems (Liza et al. 2022). It created a global shortage of various drugs required to treat the symptoms of many infections (Cundell et al. 2020). Along with that, the lockdown imposed on nations had a significant impact on the transportation of drugs (Hisham 2021). Similarly, the shortage of Personal protective equipment (Emanuel et al. 2020) for the healthcare workers worsened the situation in the healthcare sector. Sharma et al. (2020) suggested that supply chain risk management with guidelines should be adopted by the firms.

Natural disasters like earthquakes and hurricanes can cause disruptions in pharmaceutical supply chains which worsen the overall consequences. On September 20, 2017, Hurricane Maria affected Puerto Rico Island resulting in the destruction of infrastructure as well as the services including electricity and unavailability of items (Kishore et al. 2018). The Hurricane resulted in the exposure of the Pharmaceutical supply chain to Natural hazard risks. Lawrence et al. (2020a, b) investigated the major causes of this disaster in the U.S. Pharmaceutical Supply Chain by employing the Bayesian model.

The case study of Atra Pharmaceutical company (Iran) was used by Sabouhi et al. (2018) to validate the results of his approach to evaluate the resiliency of the Pharmaceutical Supply chain. The company operated in five provinces of Iran and procured raw materials from nine suppliers. It constantly maintained contacts with the potential suppliers to ensure the availability of quality raw materials for the manufacturing of drugs. To evaluate the quality of raw materials from the suppliers, Atra considered four criteria; technological compliance, after-sales response, customer satisfaction, and the fraction of low-quality material received.

Components of Pharmaceutical Supply Chain:

According to Shah (2005), a Pharmaceutical Supply Chain is defined as the management of product supply which includes the players, information, and resources dealing with the transfer of raw materials and Active Ingredients (AIs) to finished products or services and their delivery to the patient. Thus, the Suppliers, Intermediaries, and Third-Party Service Providers deal with all the related activities across the product lifecycle including clinical supply, scale-up, and transfer as well as outsourcing and product discontinuation (Pedroso and Nakano 2009). This also includes activities associated with Marketing, Sales, Product Design, Finance, and Information Technology (Secchi and Veronesi 2006).

According to Shah (2004), a typical pharmaceutical supply chain will consist of one or more of the following nodes:

i. Primary Manufacturers;
ii. Secondary Manufacturers;

iii. Distribution Centers;
iv. Wholesalers;
v. Retailers/Hospitals & Clinics;
vi. Patients.

On examining the route through which essential pharmaceutical products travel across the entire Pharmaceutical Supply Chain, the network can be categorized into three major "Stages" or "Levels."

i. Primary Stage

The Primary Stage includes the players and processes responsible for the production of the Active Pharmaceutical Ingredient (API or AI). The production usually takes place in multiple rounds of shifts, thereby, inducing a risk of delays in the production process since considerable inventories are often held between stages (Shah 2005).

ii. Secondary Stage

The Secondary Stage deals with the preparation and formulation of the products in a suitable form for final consumers. At this stage, transportation cost minimization and tax optimization become significant concerns since the Secondary sites are often geographically separate from AI producers (Cundell et al. 2020; Sousa et al. 2011).

iii. Tertiary Stage

The Tertiary Stage can be further divided into two Sub-Stages.

- In the first sub-stage, the packaging of the final product takes place and is further divided into three significant packaging types (Hosseini-Motlagh et al. 2020; Savage et al. 2006):
    i. Protection of the Drug Product Environment; and
    ii. Ensuring Product identification.
- The second sub-stage deals with the transportation of the packaged product to the wholesalers, retailers, hospitals, clinics, and eventually to the patients.

Thus, the Tertiary Stage acts as a link between the Suppliers, Intermediaries, Third-Party Service Providers, and the End Users.

According to Nair and Reed-Tsochas (2019), pharmaceutical supply chains are open systems that interact with their environment, and these interactions develop resilient strategies for the pharmaceutical supply chains. The pharmaceutical supply chains consist of production, distribution, application, research, and development units along with all the facilities, helping the departments for smooth functioning of the entire supply chains (Lawrence et al. 2020a, b; Hulbert et al. 2008; Jaberidoost et al. 2015).

2 Literature Review

According to the "Encyclopedia of Occupational Health and Safety" published by the International Labour Organisation (2011), Drugs are "substances with active pharmacological properties in humans and animals."

Drugs available from pharmaceutical sources, i.e., manufactured by the pharmaceutical industry or made up by a pharmacist are known as Pharmaceutical Drugs. Pharmaceutical Drugs can essentially be classified as follows:

- Prescription drugs or Ethical drugs—drugs that are dispensed only by prescription or approval of a medical, pharmacy, or veterinary professionals, and
- Generic Drugs or Over-The-Counter (OTC) Drugs—drugs sold in a retail store or pharmacy which do not require a prescription or the approval of a medical, pharmacy, or veterinary professional.

According to Mehralian et al. (2012), the Pharmaceutical Industry is defined as a system of processes, operations, and organizations involved in the discovery, development, and production of drugs and medications. The Research & Development as well as the Manufacture of drugs is a complex process involving many players, each having a unique characteristic and role. These key players in the pharmaceutical industry include the following (Shah 2004; He et al. 2021):

1. The Large Pharmaceutical Companies which manufacture both Generic and Prescription Drugs worldwide. These companies also invest heavily in their Research and Development.
2. Small-scale Pharmaceutical Companies that have negligible or nil Research and Development capacities. Their operations are more local and produce both out-of-patent ethical products and over-the-counter products under license or contract.
3. Contract manufacturers which provide outsourcing services to the Large Pharmaceutical companies. They produce mainly key intermediates and active ingredients (AI). These manufacturers, generally, do not have a product portfolio of their own.
4. Extremely Small-Scale Companies which have no significant manufacturing capacity. These companies work largely in the field of Drug Discovery and Biotechnology.

2.1 Identification of Risks in the Pharmaceutical Supply Chain

There has been a great amount of research on managing risks (Tang 2006); but that associated with identifying risks has a long way to go (Rao and Goldsby 2009). Some papers though have done remarkable work in identifying these risks and separating them like the work of (Rao and Goldsby 2009). However, the risks have to be identified according to their "severity of harm." Though this has been a new area, some of the papers like the work of Mehralian et al. (2012) have used various methodologies to solve this problem. Rossetti et al. (2011) describe the highly erratic nature of the pharmaceutical supply chain. They point out certain risks in the wholesale sector like the forecasting demand risk, credit risk, etc. Paulsson (2004) divided the risks into three categories: operational disturbance, tactical disturbance, and strategic uncertainty. Many factors like a sudden change in the socio-politico factors of the region, sudden inflation, and terrorism are some of the factors that have been noticed by many researchers. Based on Literature Review and Expert Opinions, the various risks in the Pharmaceutical Industry have been identified as follows:

I. *Uncertainty in the quality of the Active Pharmaceutical Ingredient/Raw Material of Drugs*:
The production process of the drug includes inputs such as material, power, labor, equipment, etc. The failure of the availability of any of these would affect the industry adversely. A drug is composed of two components—Active Pharmaceutical Ingredients (API) and Excipients. An Active Pharmaceutical Ingredient (API) refers to a biologically active central substance or substance combination used in manufacturing a drug product that is meant to produce the desired effect in the body. The World Health Organization (1994) defines an API as follows:

> Any substance or combination of substances used in a finished pharmaceutical product (FPP), intended to furnish pharmacological activity or to otherwise have direct effect in the diagnosis, cure, mitigation, treatment or prevention of disease, or to have direct effect in restoring, correcting or modifying physiological functions in human beings.

Excipients are the pharmaceutically inert substances present in the medication. These include dyes, flavors, binders, emollients, fillers, lubricants, and preservatives that make up the drug.

Hence, the quality of Active Pharmaceutical Ingredients in a drug has a direct effect on the safety and efficacy of that drug. It is, therefore, essential to ensure the high quality of these substances. Since good-quality Active Pharmaceutical Ingredients (APIs) are vital to the production of good-quality medicines, the testing of the drugs is an important stage in the production process. Improper testing may result in adverse results of the drug on mankind. Moreover, the storage, packaging, labeling, repackaging, release, and production of the APIs need to be carefully monitored as per the prevalent regulations (Hosseini-Motlagh et al. 2020). Gómez and España (2020) suggested conducting transport tests along with product filterability during the process of product development, as this will evaluate how the drugs will behave at the time of transportation.

Poorly manufactured or contaminated APIs can result in disastrous health outcomes such as therapeutic failure, exacerbation of the disease, resistance to medicines, and sometimes death (Cundell et al. 2020; Tang 2006; Asamoah et al. 2011; Pfhol et al. 2011). Saedi et al. (2016), Kuo et al. (2020) emphasize on the optimization techniques for backing up the medical shortages. Tucker et al. (2020) use this technique to study the resilience of the pharmaceutical supply chain in the US targeting low-cost medicines.

Yaroson et al. (2021) find that the timing and quality of the information shared regarding the medicines in a pharmaceutical supply chain act as a make-or-break situation. Timely, Accurate, and Purposeful information increases the supply chain visibility and provides a reliable description of demand and supply. It further helps the organization build a resilient pharmaceutical supply chain (Lawrence et al. 2020a, b).

II. *Regulatory risk*:

This Risk includes the problems caused by new or existing regulations and the uncertainty in the policies framed by the government of the region. The regulatory authorities have a large say in determining the patent life of the drugs which in turn affects the demand and price of the drug (Shah (2004). Over the last two decades, negotiations of the World Trade Organization such as the TRIPS Agreement (1994) has revolved around the issues related to the intersection of international trade in pharmaceuticals and intellectual property rights. The regulations in the developed nations generally tend to protect the intellectual property rights of the manufacturers to protect the huge investments made to develop new drugs whereas developing nations generally seek to promote their generic pharmaceuticals industries and their ability to make medicine available to their people via compulsory licenses.

Excessive patents encourage the manufacturers to invest more in the Research and Development of a Drug. This policy is useful for those drugs whose Manufacture is a highly complicated and expensive process, such as the drugs which treat Orphan diseases (diseases that affect only a small percentage of the population), and also leads to the increase in the number of Blockbuster Drugs (an extremely popular drug that generates annual sales of at least $1 billion for the company that creates it) in the market. However, this also leads to an increase in the price of Essential Drugs. On the other hand, avoiding the grant of frivolous patents and ensuring that only those products proving "substantial human intervention" and "utility" are provided with the patent protection will decrease the prices of Essential Drugs. However, it also discourages the manufacturers from investing more inthe production process.

III. *Political risk*:

The risk includes the danger of a change of government which might affect the prevalent trading laws and market conditions. The regulatory changes and the associated political fallout that accompany the change in the Political conditions in the Economy have one of the biggest impacts on the pharmaceutical companies, especially the Major Players. Since Policy Change is a profoundly political process, implementation of any Policy change in the Health Sector will have a direct impact on the prevailing market conditions and the ease of doing business. Moreover, the political conditions have a direct relationship with the Regulatory Framework (Schildhouse 2006; Ellison and

Wolfram 2006). Palit and Bhogal (2022) in their research represented significant findings on the effect of political factors in re-establishing the Indian Pharmaceutical supply chain after the COVID-19 pandemic.

IV. *Uncertainty in the Social Conditions of the region*:
This risk is concerned with terrorism, crime, employee unrest, and physical security that not only disrupt the supply chain but also affect the formulation of policies (Sheffi and Rice 2005; Freeman 1984; Lysons and Farrington 2006).

V. *Uncertainty in the Economic Activities and Prices in the region*:
This risk includes market risk, e.g., the risk that the market value of assets will fall. Pharmaceutical pricing is an important and contentious issue, especially in low- and middle-income nations. The Price of a drug depends on several factors, such as a spurt in price rise of the raw materials, labor, and inflation, which would, in turn, affect the supply chain adversely. The pricing of a drug heavily impacts its potential success. Higher price often tends to deter physicians and patients from prescribing or reimbursing it, especially in the case of Generic Drugs. This is because the benefits of the drug seem lesser in comparison to the added cost (Goodarzian et al. 2020). However, a very low price may also deter a drug's demand because the lower price might imply inferiority (Galizzi, Ghislandi, and Miraldo 2011; Kotler et al. 2002). (Kharisma and Ardi 2020) evaluated the five risks related to the generic medicines supply chain in Indonesia, and they enlisted that the ineffective production capacity, poorly formulated financial plan, inadequate production process, and escalation in the price of raw material are the top five risks.

VI. *Natural Hazard risk*:
Natural hazards such as hurricanes, earthquakes, and COVID-19 can suddenly disrupt the supply chain and affect the industry badly as they can profoundly affect drug product availability. Natural disasters cause severe damage to manufacturing facilities resulting in Long-term drug shortages. Natural Disasters also create an unexpected demand for drugs needed to treat disaster victims which further increases the shortage of drugs. Moreover, Natural Disasters may lead to the spread of epidemics which shall further put severe pressure on the supply chain (Lawrence et al. 2020a, b; Kuo et al. 2020; Ventola 2011; Watson, Gayer and Connolly 2007; Dimiturk 2005).

VII. *IT risk*:
Critical data for a Pharmaceutical Company includes product formulas, manufacturing processes, marketing plans, clinical trial data, and test results. Other important data may include a list of distribution centers, market surveys, competitor information, etc. Therefore, Loss of data, outage of the data center, etc. can affect the whole management of the supply chain right from its production process to the distribution to the customers.

Lysons and Farrington (2006); Pfhol et al. (2011). The permissioned blockchain network embedded system proposed by (Babu et al. 2022) facilitates secure information sharing across the chain and provides storage reliability of Supply Drugs Information (SDI). Also, the technologies including the Internet of Things (IoT) can be applied in the pharmaceutical supply chain to lower the risk factors associated with it (Pargaien et al. 2022).

VIII. *Prevalence of Counterfeit Drugs*:
According to the World Health Organization (1994),

> A counterfeit medicine is one which is deliberately and fraudulently mislabeled with respect to identity and/or source. Counterfeiting can apply to both branded and generic products and counterfeit products may include products with the correct ingredients or with the wrong ingredients, without active ingredients, with insufficient active ingredients, or with fake packaging.

In addition, legitimate drugs are sold again after their date of expiry by remarking it with a false date. A counterfeit drug is unlikely to be suspicious, especially if it comes from a supposedly legitimate source. Moreover, unlike other consumer goods, the consumer's own judgment of a product is very less in the case of pharmaceutical drugs since they are usually prescribed by a doctor or a health worker. Even when patients choose their own drugs they may lack the specialized knowledge to detect whether the product they are buying is of good quality, thereby, making it extremely difficult to distinguish between a legitimate and a counterfeit drug.

Manufacture of Counterfeit Drugs is a relatively cheaper process since they do not require building huge infrastructure or facilities. Moreover, there are also no overhead costs due to quality assurance or meeting Good Manufacturing Practices (GMP) standards (Cundell et al. 2020). These costs may further be decreased if cheap substitutes are used or if these are omitted altogether, as is often the case. Low-quality counterfeit medication may cause several dangerous health consequences, including side effects or allergic reactions.

The entry of counterfeit drugs into the market has been quite a challenging risk for the existing pharmaceutical drugs which decreases their demand and therefore this leads to an uncertainty in the demand for the drugs (Shah 2004; Cross 2007). Stevenson and Busby (2015) highlighted counterfeiting as a major problem in pharmaceutical supply chains. Counterfeiting directly threatens consumer welfare and has negative social and economic impacts.

Knowledge Transfer is considered as an important parameter in counterfeiting. Das et al. (2006) found that undesirable knowledge transfer between the suppliers leads to counterfeiting. Organizations must be sure that no external stakeholders get the complete information about the product's process or product (e.g., Van Hoek and Weken 1998; Jacobs et al. 2007). Captive offshoring and off-shore outscoring hold a major linkage with the counterfeiting threat (Stevenson and Busby 2015; Ellram et al. 2008; Enyinda et al. 2009; Mokrini et al. 2016a) Mokrini et al. 2016b). One of the reasons for Captive offshoring is knowledge sharing. New technologies have proven to protect pharmaceutical supply chains from counterfeiting. For example, the RFID tags have been observed providing the traceability of the products, helping the organizations to overcome counterfeiting (Visich et al. 2009;

Maruckeck et al. 2011). Also, the study by Bhansali et al. (2021) suggests that the implementation of blockchain enhances pharmaceutical supply chain management and lowers the risk of counterfeit, unauthorized drugs, and related equipment.

IX. *Risk of Losing Key Suppliers*:
The major sources of supply chains are its suppliers (Liao et al. 2021). The selection of reliable suppliers results in an increase in market share and helps the firms to gain a competitive edge (Praveen and Suraj 2021). The risk of losing key suppliers can be worse if there is an overdependence on them. Losing them may be like losing a stronghold on the market (Pfhol et al. (2011)]. To achieve sustainable enterprise management, it is essential to form good relationships with suppliers (Modibbo et al. 2022).

X. *Logistics and Inventory Risk*:
Pharmaceutical players need to maintain optimum inventory levels and efficient distribution networks to meet patient demand on time and prevent excessive inventory since perishable pharmaceutical drugs go to waste if not kept under strictly controlled environments. Thus, there is logistics and inventory control over the supply of drugs to foresee a correct demand in the future, expiry of products before sale, etc. Mismanagement of Logistics and Inventory can affect the industry adversely (Asamoah et al. 2011; Tang 2006; Shah 2004). The management of Inventory is essential as it may result in a lowered cost incurred in the process of manufacturing Pharmaceutical products (Padmavathi and Rajagopalan 2022).

Several researchers have contributed to the assessment of risks involved in the field of Pharmaceutical Supply chains using various multi-criteria decision-making approaches. However, the literature is lacking such efforts in Indian perspective. For instance, the work carried out by (Jlassi et al. 2021) identified the major risks in the pharmaceutical supply chain in Tunisia during the COVID-19 pandemic using the fuzzy AHP method. Raka and Liangrokapart (2017) in their study in Thailand focused on the risks incurred during the development process of a new drug and further prioritized them using AHP. Jaberidoost et al. (2015) used the AHP and SAW methods to analyze risks in their quantitative study in Iran. The majority of researches in India are focused on the evaluation of barriers and enablers in the Pharmaceutical supply chain. Chandra et al. (2021) examined the significant factors responsible for enhancing the overall performance of the existing vaccine supply chain in India by applying the next-generation vaccine supply chain. Sharma and Joshi (2021) investigated the barriers to the implementation of blockchain in the Indian healthcare sector. In this paper, the Technique for Order of Preference by Similarity to Ideal Solution (TOPSIS) is utilized in an effort to prioritize the risks related to the Pharmaceutical Industry in India.

3 Research Methodology

In this study for ranking different risks, the Technique for Order Preference by Similarity to Ideal Solution (TOPSIS) approach has been used. TOPSIS is a multi-criteria decision analysis method that enables the Decision Makers (DMs) to rank the criteria based on alternatives. TOPSIS selects the alternative that is closest to the ideal solution and farthest from the negative ideal solution. Accordingly, the selection of a suitable alternative(s) is made (Shih et al. 2007). Different steps of TOPSIS are given below.

I. *Step 1*:
   Let there be $M = 3$ alternatives to each of the $N = 10$ criteria. The weights are then assigned to each alternative.
   Then Construct a Decision Matrix where X_{ij} shows the rating of its *i*th alternative concerning the *j*th criteria.
   Where i = 1, 2, 3 is the number of the alternative and $j = 1, 2, 3, \ldots,10$ is the number of criteria.
II. *Step 2*:
   Construct the Normalized Decision Matrix as
   $R = r_{ij}$ where

$$r_{ij} = \frac{X_{ij}}{\sqrt{\sum_{i=1}^{M} X_{ij}^2}} \tag{1}$$

   and R is the normalized matrix of element r_{ij}.
III. *Step 3*:
   Construct the Weighted Normalized Matrix by multiplying the elements of the normalized matrix by the weights of corresponding criteria as

$$V_{ij} = r_{ij} * W_j \tag{2}$$

   where V_{ij} is the weighted normalized matrix and W_j is the weight of the *j*th criteria.
IV. *Step 4*:
   Determine the positive and negative ideal solutions as V_j^+ and V_j^-, respectively, by finding the maximum and minimum values of weighted normalized elements in each column.
V. *Step 5*:
   Calculate the separation measures for each alternative as

$$\text{the } S_{i+} = \sqrt{\sum (V_{ij} - V_{j+})^2} \tag{3}$$

   $i = 1, 2, 3$ and $j = 1, 2, 3, \ldots, 10$

And

$$S_{i-} = \sqrt{\sum (V_{ij} - V_{j-})^2} \tag{4}$$

$i = 1, 2, 3$ and $j = 1, 2, 3, \ldots, 10$

VI. *Step 6*:
Calculate the relative closeness to the ideal solution using the following formula:

$$C_{i+} = \frac{S_{i-}}{S_{i+} + S_{i-}} \tag{5}$$

$0 \leq C_{i+} \leq 1, i = 1, 2, 3$

VII. *Step 7*:
Rank the preference order as C_{i+}. The best alternative is the one that has the shortest distance to the ideal solution.

4 Results and Discussion

From the literature review, ten risks are identified. These are summarized in Table 1. For the prioritization of these risks, the TOPSIS approach has been approved. The different steps have been explained in the methodology section. The weights allotted to the three Decision Makers are shown in Table 2. After taking ratings from three Decision Makers, Normalized Decision Matrix has been constructed, as shown in Table 3. As per Eq. (2), a Weighted Normalized Decision Matrix has been constructed, shown in Table 4. The Positive and Negative ideal solutions have been determined by finding the maximum and minimum values, respectively, of weighted normalized elements in each column. The final closeness ratio of all factors is calculated as given in Table 5. Thus, based on the Relative Closeness of the Ideal Solution, the ranking of the Risks is as follows.

CD > AI > RR > LR > PR > ER > RS > IT > SR > NH

Based on TOPSIS ranking of risks, it has been observed that Risk due to Counterfeit Drugs has been given the highest priority. Thus, it is crucial for Drug manufacturers and distributors to invest heavily in the countermeasures, such as traceability and authentication technologies, to try to minimize the impact of counterfeit drugs. After Counterfeit Drugs, the factor which poses the utmost risk to the stakeholders in the supply chain is the "Uncertainty in the quality of the Active Pharmaceutical Ingredient/Raw Material of drugs." Thus, new concepts and technologies, which address the dual challenge of accelerating the drug development process as well as terminating unpromising compounds during the development process itself, need to be developed through innovation and effective decision-making. However, it is also essential to validate the effectiveness of these measures before adopting them.

Table 1 Summary of identified risks (Created by Authors)

S.No	Risks	Symbol	References
1	Uncertainty in the quality of the active pharmaceutical ingredient/raw material of drugs	(UQ)	Hosseini-Motlagh et al. (2020), Tang (2006), Asamoah et al. (2011), Pfhol et al. (2011)
2	Regulatory risks	*(RR)*	Shah (2004)
3	Political risk	*(PR)*	Schildhouse (2006), Ellison and Wolfram (2006)
4	Uncertainty in the social conditions of the region	*(SR)*	Sheffi and Rice(2005), Freeman (1984), Lysons and Farrington (2006)
5	Uncertainty in the economic activities and prices in the region	*(ER)*	Danzon et al. (2013), Galizzi et al. (2011), Kotler et al. (2002)
6	Natural hazards	*(NH)*	Ventola (2011), Watson, Gayer and Connolly (2007), Dimiturk (2005)
7	IT breakdowns	*(IT)*	Lysons and Farrington (2006), Pfhol et al. (2011)
8	Prevalence of counterfeit drugs	*(CD)*	Shah (2004), Cross (2007)
9	Loss of key suppliers	*(RS)*	Pfhol et al. (2011)
10	Logistics and inventory risk	*(LR)*	Asamoah et al. (2011), Tang (2006), Shah (2004)

Table 2 Weights for the decision variables (Created by Authors)

	D1	D2	D3
Weights	0.27	0.4	0.33

Table 3 Normalized decision matrix (Created by Authors)

	D1	D2	D3
UQ	0.43	0.37	0.36
RR	0.34	0.47	0.27
PR	0.26	0.28	0.36
SR	0.26	0.28	0.27
ER	0.34	0.28	0.27
NH	0.17	0.28	0.27
IT	0.34	0.28	0.18
CD	0.34	0.37	0.45
RS	0.26	0.35	0.27
LR	0.34	0.28	0.36

The Regulatory Risks, Logistics and Inventory Risks, and the Political Risks occupy the third, fourth, and fifth positions, respectively, in the priority order. The causes and effects of these factors are interrelated and have spontaneous implications on the pharmaceutical supply chain. It is, therefore, necessary for the Public Policy

Table 4 Weighted normalized decision matrix (Created by Authors)

	D1	D2	D3
UQ	0.12	0.15	0.12
RR	0.09	0.19	0.09
PR	0.07	0.11	0.12
SR	0.07	0.11	0.09
ER	0.09	0.11	0.09
NH	0.05	0.11	0.09
IT	0.09	0.11	0.06
CD	0.09	0.15	0.15
RS	0.07	0.14	0.09
LR	0.09	0.11	0.12

Table 5 Closeness ratio of risks (Created by Authors)

Risks	Closeness ratio	Rank
UQ	0.71	2
RR	0.66	3
PR	0.45	5
SR	0.33	9
ER	0.40	6
NH	0.29	10
IT	0.33	8
CD	0.74	1
RS	0.38	7
LR	0.50	4

Makers, Regulators, Manufacturers, and the Distributors to work together in key parts of the business. They need to be connected in different ways and empowered to develop healthy and fresh approaches to the business models. Moreover, the Policy Makers need to take steps to make the domestic industry more robust and create an environment that is conducive to research. Patent Policies that encourage the development of innovative new drugs through both R&D and Discovery Research within the industry need to be established. Encouraging Reverse Engineering will also play a crucial role in addressing these risks. However, the regulations monitoring these innovations need to be put in place to safeguard the needs and to ensure the safety of the consumers.

The sixth most important factor is the risk due to Economic Uncertainties. To address this, the Policy Makers and the Regulators need to encourage the Manufacturers and the Distributors through monetary incentives. The provision of greater subsidies will allow companies to make additional profits which will in turn allow them to invest more in Research. It is also essential to provide pharmaceutical companies with greater tax incentives which are necessary to attract greater foreign investment into a country. The seventh and eighth most important risks have been identified as the loss of key suppliers and IT breakdowns, respectively, whereas the risks due to the uncertainties in the social conditions and natural hazards occupy the ninth and tenth position, respectively. To address all of these factors, there need to be increased partnerships between the governments and the industry stakeholders, and measures need to be devised to ensure people's access to the drugs they need. To facilitate such measures, suitable infrastructure needs to be put in place and outsourcing activities need to be encouraged.

The findings of the study will help managers in handling different risks coming into supply chains. Moreover, there needs to be a substantial increase in healthcare spending around the world as per the Policy and Regulatory Framework. Thus, through increased emphasis on Innovation and Research and greater coordination among the various stakeholders, all these risks can be mitigated effectively and the efficiency of the Pharmaceutical Supply Chain can be increased.

5 Conclusion, Limitations, and Future Scope

In this study, objective was to identify and prioritize major risks in pharmaceutical supply chains. Study identified ten risks and these are prioritized by applying the TOPSIS approach. Risk due to Counterfeit Drugs has got the highest priority, highlighting the need for new technological advancements to prevent this risk. These advancements can maintain the authenticated network of reliable suppliers in the pharmaceutical supply chain. Therefore, management should adopt necessary initiatives to create awareness among the individuals about this risk to lower its outcomes. The second highest priority is given to the risk of uncertainty in the quality of the active pharmaceutical ingredient/ raw materials of drugs, suggesting the importance of a reliable drug development process. The processes involved in the pharmaceutical supply chain have a major impact on the quality of medicines and eventually on the health of patients. Next in order are the Regulatory Risks, Logistics and Inventory risks, and the political risks.

The findings of the study will be of significant value for pharmaceutical industry managers in formulating effective strategies to be resilient and sustainable. However, findings cannot be generalized because of some limitations due to the biased opinion of experts. Therefore, results may be validated with other techniques and case studies as a future scope for study.

References

Asamoah D, Abor P, Opare M (2011) An examination of pharmaceutical supply chain for artemisinin-based combination therapies in Ghana. Manag Res Rev 34(7):790–809

Babu ES, Kavati I, Nayak SR, Ghosh U, Al Numay W (2022) Secure and transparent pharmaceutical supply chain using permissioned blockchain network. Int J Logist Res Appl 1–28. https://doi.org/10.1080/13675567.2022.2045578

Barroso AP, Machado VH, Machado VC (2010) The resilience paradigm in the supply chain management: a case study. In: Proceedings of IEEE international conference on industrial engineering and engineering management

Baryannis G, Validi S, Dani S, Antoniou G (2019) Supply chain risk management and artificial intelligence: state of the art and future research directions. Int J Prod Res 57(7):2179–2202

Bhansali A, Masih J, Sharma M (2021) Blockchain 3.0 for sustainable healthcare. In: Blockchain 3.0 for sustainable development, pp 101–122

Bigdeli M, Jacobs B, Tomson G, Laing R, Ghaffar A, Dujardin B, Damme W (2013) Access to medicines from a health system perspective. Health Policy Plan 28(7):692–704. https://doi.org/10.1093/heapol/czs108

Blackhurst J, Dunn KS, Craighead CW (2011) An empirically derived framework of global supply resiliency. J Bus Logist 32(4):374–391

Breen L (2008) A preliminary examination of risk in the pharmaceutical supply chain (PSC) in the National Health Service (NHS) (UK). J Serv Sci Manag 1(2):193–99

Brindley CS (ed) (2004) Supply chain risk. Ashgate Publishing, Aldershot

Chandra D, Vipin B, Kumar D (2021) A fuzzy multi-criteria framework to identify barriers and enablers of the next-generation vaccine supply chain. Int J Product Perform Manag. https://doi.org/10.1108/IJPPM-08-2020-0419

Chopra S, Meindl P (2014) Supply chain management—Strategy, planning, and operation. https://doi.org/10.1007/s13398-014-0173-7.2

Christopher M, Lee H (2004) Mitigating supply chain risk through improved confidence. Int J Phys Distrib Logist Manag 34(5):388–396. https://doi.org/10.1108/09600030410545436

Council of Supply Chain Management Professionals (2013) Supply chain management—Terms and glossary

Craighead CW, Blackhurst J, Rungtusanatham MJ, Handfield RB (2007) The severity of supply chain disruptions: design characteristics and mitigation capabilities. Decis Sci 38(1):131–156. https://doi.org/10.1111/j.1540-5915.2007.00151.x

Cross C (2007) A hard sell: Carolina logistics services reverses the route of controlled substances. Ind Eng 39(5):24

Cundell T, Guilfoyle DE, Kreil TR, Sawant A (2020) Controls to minimize disruption of the pharmaceutical supply chain during the COVID-19 pandemic. PDA J Pharm Sci Technol 74:468–494

Danzon PM, Mulcahy AW, Towse AK (2013) Pharmaceutical pricing in emerging markets: effects

Daghfous A, Qazi A, Khan MS (2021) Incorporating the risk of knowledge loss in supply chain risk management. Int J Logist Manag 32(4):1384–1405. https://doi.org/10.1108/IJLM-06-2020-0225

Das A, Narasimhan R, Talluri S (2006) Supplier integration-finding an optimal configuration. J Oper Manag 24(5):563–582

Derrong Lin I, Hertig JB (2021) Pharmacy leadership amid the pandemic: maintaining patient safety during uncertain times. Hospital Pharm 57(3):323–328

Dimiturk A (2005) Three keys to supply chain management in times of disaster. Healthc Purch News 29(12):64–65

Dubey R, Gunasekaran A, Bryde DJ, Dwivedi YK, Papadopoulos T (2020) Blockchain technology for enhancing swift-trust, collaboration and resilience within a humanitarian supply chain setting. Int J Prod Res 58(11):3381–3398. https://doi.org/10.1080/00207543.2020.1722860

Dubey R, Gunasekaran A, Childe SJ, Papadopoulos T, Blome C, Luo Z (2019) Antecedents of resilient supply chains: an empirical study. IEEE Trans Eng Manag 66(1):8–18
El Baz J, Ruel S (2021) Can supply chain risk management practices mitigate the disruption impacts on supply chains' resilience and robustness? Evidence from an empirical survey in a COVID-19 outbreak era. Int J Prod Econ 233:107972. https://doi.org/10.1016/j.ijpe.2020.107972
Elleuch H, Hachicha W, Chabchoub H (2013) A combined approach for supply chain risk management: description and application to a real hospital pharmaceutical case study. J Risk Res 17(5):641–663. https://doi.org/10.1080/13669877.2013.815653
Ellison SF, Wolfram C (2006) Coordinating on lower prices: pharmaceutical pricing under political pressure. RAND J Econ 37(2):324–340
Ellram LM, Tate WL, Billington C (2008) Offshore outsourcing of professional services: a transaction cost economics perspective. J Oper Manag 26(2):148–163
Emanuel EJ, Persad G, Upshur R, Thome B, Parker M, Glickman A, Zhang C, Boyle C, Smith M, Phillips JP (2020) Fair allocation of scarce medical resources in the time of Covid-19. N Engl J Med 382(21):2049–2055. https://doi.org/10.1056/NEJMsb2005114
Enyinda C, Briggs C, Bachkar K (2009) Managing risk in pharmaceutical global supply chain outsourcing: applying analytic hierarchy process model. Am Soc Bus Behav Sci 16:19–22
Finch P (2004) Supply chain risk management. Supply Chain Manag: Int J 9(2):183–196. https://doi.org/10.1108/13598540410527079
Freeman RE (1984) Strategic management: a stakeholder approach. Pitman, Boston, MA
Galizzi MM, Ghislandi S, Miraldo M (2011) Effects of reference pricing in pharmaceutical markets: a review. Pharmacoeconomics 29(1):17–33
Gandhi S, Mangla SK, Kumar P, Kumar D (2016) A combined approach using AHP and DEMATEL for evaluating success factors in implementation of green supply chain management in Indian manufacturing industries. Int J Logist Res Appl 19(6):537–561. https://doi.org/10.1080/13675567.2016.1164126
Giannakis M, Croom S, Slack N (2004) Supply chain paradigms. In: New S, Westbrook R (eds) Understanding supply chains. Oxford University Press, Oxford, pp 1–22
Goodarzian F, Nasab HH, Muñuzuri J, Fakhrzad MB (2020) A multi-objective pharmaceutical supply chain network based on a robust fuzzy model: a comparison of meta-heuristics. Appl Soft Comput 92:106331
Gómez JCO, España KT (2020) Operational risk management in the pharmaceutical supply chain using ontologies and fuzzy QFD. Procedia Manuf 51:1673–1679. https://doi.org/10.1016/j.promfg.2020.10.233
He X, Liu W, Hu W, Wu X (2021) Coalitional strategies of the pharmaceutical supply chain with an option contract to cope with disruption risks. Int J Logist Res Appl
Hisham A (2021) Global drug shortages due to COVID-19: impact on patient care and mitigation strategies. Res Social Adm Pharm 17(1):1946–1949. https://doi.org/10.1016/j.sapharm.2020.05.017
Hohenstein N-O (2022) Supply chain risk management in the COVID-19 pandemic: strategies and empirical lessons for improving global logistics service providers' performance. Int J Logist Manag. https://doi.org/10.1108/IJLM-02-2021-0109
Hosseini-Motlagh S, Jazinaninejad M, Nami N (2020) Recall management in pharmaceutical industry through supply chain coordination. Ann Oper Res 1–39
Hulbert MH, Feely LC, Inman EL, Johnson AD, Kearney AS, Michaels J, Mitchell M, Zour E (2008) Risk management in the pharmaceutical product development process. J Pharm Innov 3(4):227–248. https://doi.org/10.1007/s12247-008-9049-8
Hung Y, Huang S, Lin Q, Tsai M (2005) Critical factors in adopting a knowledge management system for the pharmaceutical industry. Ind Manag Data Syst 105(2):164–183. https://doi.org/10.1108/02635570510583307
International Labour Organization (2011) Encyclopedia of occupational health and safety
International Organization for Standardization. ISO 31000:2009 Risk management—Principles and guidelines

Ivanov D, Dolgui A (2020) A digital supply chain twin for managing the disruption risks and resilience in the era of Industry 4.0. Prod Plan Control 58. 10.1080/ 09537287.2020.1768450

Jaberidoost M, Nikfar S, Abdollahiasl A, Dinarvand R (2013) Pharmaceutical supply chain risks: a systematic review. Daru 21(69):1–7. https://doi.org/10.1186/2008-2231-21-69

Jaberidoost M, Olfat L, Hosseini A, Kebriaeezadeh A, Abdollahi M, Alaeddini M, Dinarvand R (2015) Pharmaceutical supply chain risk assessment in Iran using analytic hierarchy process (AHP) and simple additive weighting (SAW) methods. J Pharm Policy Pract 8(1):1–10. https://doi.org/10.1186/s40545-015-0029-3

Jacobs M, Vickery SK, Droge C (2007) The effects of product modularity on competitive performance: do integration strategies mediate the relationship? Int J Oper Prod Manag 27(10):1046–1068

Jlassi J, Halouani N, El Mhamedi A (2021) Pharmaceutical supply chain risk assessment in the time of COVID 19/case study. In: Dolgui A, Bernard A, Lemoine D, von Cieminski G, Romero D (eds) Advances in production management systems. artificial intelligence for sustainable and resilient production systems. Springer International Publishing, pp 697–704. https://doi.org/10.1007/978-3-030-85906-0_76

Kamalahmadi M, Parast MM (2016) A review of the literature on the principles of enterprise and supply chain resilience: major findings and directions for future research. Int J Prod Econ 171:116–133

Kersten W, Böger M, Hohrath P, Späth H (2006) Supply chain risk management: development of a theoretical and empirical framework. In: Kersten W, Blecker T (eds) Managing risks in supply chains: how to build reliable collaboration in logistics, Erich Schmidt, Berlin, pp 3–18

Khan O, Burnes B (2007) Risk and supply chain management: creating a research agenda. Int J Logist Manag 18(2):197–216

Kharisma SA, Ardi R (2020) Supply chain risk assessment of generic medicine in Indonesia using DEMATEL-based ANP (DANP). In: IEEE international conference on industrial engineering and engineering management, 2020-Decem, pp 716–720. https://doi.org/10.1109/IEEM45057.2020.9309793

Kilubi I (2016) The strategies of supply chain risk management—A synthesis and classification. Int J Log Res Appl 19(6):604–629

Kishore N, Marqués D, Mahmud A, Kiang MV, Rodriguez I, Fuller A, Ebner P, Sorensen C, Racy F, Lemery J, Maas L, Leaning J, Irizarry RA, Balsari S, Buckee CO (2018) Mortality in Puerto Rico after Hurricane Maria. N Engl J Med 379(2):162–170. https://doi.org/10.1056/NEJMsa1803972

Kotler P, Armstrong G, Saunders J, Wong V (2002) Principles of marketing, 3rd ed. Pearson Education, Harlow

Kumar P (2008) An integrated model of AHP and TOPSIS for 3PL evaluation. Asia-Pacific Bus Rev 4(3):14–21

Kuo S, Ou H, Wang CJ (2020) Managing medication supply chains: lessons learned from Taiwan during the COVID-19 pandemic and preparedness planning for the future. J Am Pharm Assoc 61:e12–e15

Kwak Y, Dixon CK (2008) Risk management framework for pharmaceutical research and development projects. Int J Manag Proj Bus 1(4):552–565. https://doi.org/10.1108/17538370810906255

Lawrence J, Ibne Hossain NU, Jaradat RM, Hamilton M (2020a) Leveraging a Bayesian network approach to model and analyze supplier vulnerability to severe weather risk: a case study of the U.S. pharmaceutical supply chain following Hurricane Maria. Int J Disaster Risk Reduct 49:101607–101607

Lawrence J, Ullah N, Hossain I, Jaradat R, Hamilton M (2020b) Leveraging a Bayesian network approach to model and analyze supplier vulnerability to severe weather risk: a case study of the U. S. pharmaceutical supply chain following Hurricane Maria. Int J Disast Risk Reduct 49(August 2019):101607. https://doi.org/10.1016/j.ijdrr.2020.101607

Li H, Liu S-M, Yu X-H, Tang S-L, Tang C-K (2020a) Coronavirus disease 2019 (COVID-19): current status and future perspectives. Int J Antimicrob Agents 55(5):105951. https://doi.org/10.1016/j.ijantimicag.2020.105951

Li Y, Zobel C, Seref O, Chatfield D (2020b) Network characteristics and supply chain resilience under conditions of risk propagation. Int J Prod Econ 223:1–13

Liao H, Kuang L, Liu Y, Tang M (2021) Non-cooperative behavior management in group decision making by a conflict resolution process and its implementation for pharmaceutical supplier selection. Inf Sci 567:131–145. https://doi.org/10.1016/j.ins.2021.03.010

Liza SA, Chowdhury NR, Paul SK, Morshed M, Morshed SM, Bhuiyan MAT, Rahim MdA (2022) Barriers to achieving sustainability in pharmaceutical supply chains in the post-COVID-19 era. Int J Emerg Mark. https://doi.org/10.1108/IJOEM-11-2021-1680

Luthra S, Mangla SK, Venkatesh K, Jakhar S (2017) Prioritization and management of risks in sustainable supply chain using AHP and Monte Carlo simulation. In: Managerial Strategies for Business Sustainability during Turbulent Times, publisher, IGI Global, Hershey, PA

Lysons K, Farrington B (2006) Purchasing and supply chain management, 7th edn. Financial Times/ Prentice-Hall, London

Mahendran H, Narasimhan K, Nagarajan N, Gopinath S (2011) Investigation of supply chain risk in the Indian pharmaceutical industry: a case study. In: Proceedings of World Congress on engineering, pp 836–841

Mangla SK, Kumar P, Barua MK (2015) Risk analysis in green supply chain using fuzzy AHP approach: a case study. Resour Conserv Recycl 104:375–390. https://doi.org/10.1016/j.resconrec.2015.01.001

Mangla SK, Kumar P, Barua MK (2016) An integrated methodology of FTA and fuzzy AHP for risk assessment in green supply chain. Int J Oper Res 25(1):77–99. https://doi.org/10.1504/IJOR.2016.073252

Manuj I, Mentzer JT (2008) Global supply chain risk management strategies. Int J Phys Distrib Logist Manag 38(3):192–223. https://doi.org/10.1108/09600030810866986

Maruckeck A, Greis N, Mena C, Cai L (2011) Product safety and security in the global supply chain: issues, challenges and research opportunities. J Oper Manag 29(7):707–720

Mehralian G, Gatari AR, Morakabati M, Vatanpour H (2012) Developing a suitable model for supplier selection based on supply chain risks: an empirical study from iranian pharmaceutical companies. Iran J Pharm Res 11(1):209–219

Modgil S, Sharma S (2016) Total productive maintenance, total quality management and operational performance an empirical study of Indian pharmaceutical industry. J Qual Maint Eng 22(4):353–377

Modibbo UM, Hassan M, Ahmed A, Ali I (2022) Multi-criteria decision analysis for pharmaceutical supplier selection problem using fuzzy TOPSIS. Manag Decis 60(3):806–836. https://doi.org/10.1108/MD-10-2020-1335

Mokrini A, Dafaoui E, Berrado A, El Mhamedi A (2016a) An approach to risk assessment for outsourcing logistics: case of pharmaceutical industry. IFAC-PapersOnLine 49(12):1239–1244. https://doi.org/10.1016/j.ifacol.2016.07.681

Mokrini A, Kafa N, Dafaoui E, El Mhamedi A, Berrado A (2016b) Evaluating outsourcing risks in the pharmaceutical supply chain: case of a multi-criteria combined fuzzy AHP-PROMETHEE approach. IFAC-PapersOnLine 49(28):114–119. https://doi.org/10.1016/j.ifacol.2016.11.020

Moktadir MA, Ali SM, Mangla SK, Sharmy TA, Luthra S, Mishra N, Garza-Reyes JA (2018) Decision modeling of risks in pharmaceutical supply chains. Ind Manag Data Syst

Mouloudi L, Evrard Samuel K (2022) Critical materials assessment: a key factor for supply chain risk management. Supply Chain Forum: Int J 23(1):53–67. https://doi.org/10.1080/16258312.2021.2008771

Mustaffa NH, Potter A (2009) Healthcare supply chain management in Malaysia: a case study. Supply Chain Manag: an Int J 14(3):234–243

Nair A, Reed-Tsochas F (2019) Revisiting the complex adaptive systems paradigm: leading perspectives for researching operations and supply chain management issues. J Oper Manag 65(2):80–92

O'Connor T, Yang X, Tian G, Chatterjee S, Lee S (2016) Quality risk management for pharmaceutical manufacturing: the role of process modeling and simulations. Predict Model Pharm Unit Oper 15:15–37. https://doi.org/10.1016/B978-0-08-100154-7.00002-8

Ozdemir D, Sharma M, Dhir A, Daim T (2022) Technology in Society Supply chain resilience during the COVID-19 pandemic. Technol Soc 68(October 2021):101847. https://doi.org/10.1016/j.techsoc.2021.101847

Paulsson U (2004) Supply chain risk management. In: Brindley C (ed) Supply chain risk management. Ashgate, Aldershot, pp 79–96

Pedroso MC, Nakano D (2009) Knowledge and information flows in supply chains: a study on pharmaceutical companies. Int J Prod Econ 122:376–384

Padmavathi U, Rajagopalan N (2022) An interactive drug supply chain tracking system using 13(1):221–237

Palit A, Bhogal P (2022) COVID19, Supply Chain Resilience, and India: Prospects of the Pharmaceutical Industry. In: Palit A (ed) Globalisation impacts: countries, institutions and COVID19. Springer Singapore, pp 159–181. https://doi.org/10.1007/978-981-16-7185-2_9

Pargaien AV, Kumar T, Maan M, Sharma S, Joshi H, Pargaien S (2022) IoT: a revolutionizing step for pharmaceutical supply chain management during and after covid19. Lect Notes Netw Syst 213:349–360. https://doi.org/10.1007/978-981-16-2422-3_28

Pfhol H-C, Gallus PhiliAI, Thomas D (2011) Interpretive structural modeling of supply chain risks. Int J Phys Distrib Logist Manag 41(9):839–859

Porter M (1985) Competitive advantage. The Free Press, New York, NY

Praveen T, Suraj V (2021) A study on supplier and buyer coordination on pharma supply chain. Prayukti—J Manag Appl 01(01). https://doi.org/10.52814/PJMA.2021.1102

Privett N, Gonsalvez D (2014) The top ten global health supply chain issues: perspectives from the field. Oper Res Health Care 3(4):226–230. https://doi.org/10.1016/j.orhc.2014.09.002

Rao S, Goldsby TJ (2009) Supply chain risks: a review and typology. Int J Logist Manag 20(1):97–123

Raka C, Liangrokapart J (2017) An analytical hierarchy process (AHP) approach to risk analysis: a case study of a new generic drug development process. J Pharm Innov 12(4):319–326. https://doi.org/10.1007/s12247-017-9298-5

Reich MR (2005) The politics of health sector reform in developing countries: three cases of pharmaceutical policy. Health Policy 32:47–77

Rogachev AY (2008) Enterprise risk management in a pharmaceutical company. Risk Manag 10(1):76–84. https://doi.org/10.1057/palgrave.rm.8250037

Rossetti CL, Handfield R, Dooley KJ (2011) Forces, trends, and decisions in pharmaceutical supply chain management. Int J Phys Distrib Logist Manag 41(6):601–622

Sabouhi F, Pishvaee MS, Jabalameli MS (2018) Resilient supply chain design under operational and disruption risks considering quantity discount: a case study of pharmaceutical supply. Comput Ind Eng. https://doi.org/10.1016/j.cie.2018.10.001

Saedi S, Kundakcioglu OE, Henry AC (2016) Mitigating the impact of drug shortages for a healthcare facility: an inventory management approach. Eur J Oper Res 251(1):107–123

Savage CJ, Roberts KJ, Wang XZ (2006) A holistic analysis of pharmaceutical manufacturing and distribution: are conventional supply chain techniques appropriate? Pharm Eng Int Soc Pharm Eng 26(4)

Schildhouse J (2006) An interview with Willis D. Pugh. J Supply Chain Manag 42(1):2–3

Scholten K, Scott PS, Fynes B (2019) Building routines for non-routine events: supply chain resilience learning mechanisms and their antecedents. Supply Chain Manag Int J 24(3)

Secchi R, Veronesi V (2006) The adoption of collaborative practices: a survey on the pharmaceutical supply chain. In: 22nd IMP-conference. Industrial Marketing and Purchasing Group, Milan, Italy

Shah N (2004) Pharmaceutical supply chains: key issues and strategies for optimisation. Comput Chem Eng 28:929–941
Shah N (2005) Process industry supply chains: advances and challenge. Comput Chem Eng 29:1225–1235
Sharma S, Modgil S (2020) TQM, SCM and operational performance: an empirical study of Indian pharmaceutical industry. Bus Process Manag J 26(1):331–370
Sharma A, Gupta P, Jha R (2020) COVID-19: impact on health supply chain and lessons to be learnt. J Health Manag 22(2):248–261. https://doi.org/10.1177/0972063420935653
Sharma M, Joshi S (2021) Barriers to blockchain adoption in health-care industry: an Indian perspective. J Glob Oper Strateg Sour 14(1):134–169. https://doi.org/10.1108/JGOSS-06-2020-0026
Sheffi Y, Rice J (2005) A supply chain view of the resilient enterprise. MIT Sloan Manag Rev 47(1):41–48
Shenoi VV, Dath TNS, Rajendran C, Shahabudeen P (2018) Strategic action grids: a study on supply chain risk management in manufacturing industries in India. Benchmark: Int J 25(8):3045–3061
Shih HS, Shyur HJ, Lee ES (2007) An extension of TOPSIS for group decision making. Math Comput Model 45(7–8):801–813
Singh NP, Singh S (2019) Building supply chain risk resilience Role of big data analytics in supply chain disruption mitigation. Benchmark: Int J 26(7):2318–2342
Sodhi MS, Son B-G, Tang CS (2012) Researchers' perspectives on supply chain risk management. Prod Oper Manag 21(1):1–13
Sousa RT, Liu S, Papagiorgiou LG, Shah N (2011) Global supply chain planning for pharmaceuticals. Comput Chem Eng 89:2396–2409
Stevenson M, Busby J (2015) Towards counterfeit-resilient supply chains. Int J Oper Prod Manage 35(1):110–144
Sriyakul T, Umam R, Jermsittiparsert K (2019) Internal supply chain integration and operational performance of indonesian fashion industry firms. Humanit Soc Sci Rev 7(2):479–486. https://doi.org/10.18510/hssr.2019.7256
Tang CS (2006) Perspectives in supply chain risk management. Int J Prod Econ 103(2):451–488
Tazin F (2016) Pharmaceutical industry of Bangladesh: progress and prospects. Millenn Univ J 1:19–30. https://themillenniumuniversity.edu.bd/image/TMUJ_P19.pdf
Tucker EL, Daskin MS, Sweet BV, Hopp WJ (2020) Incentivizing resilient supply chain design to prevent drug shortages: policy analysis using two-and multistage stochastic programs. IISE Trans 52(4):394–412
Tukamuhabwa BR, Stevenson M, Busby J, Zorzini M (2015) Supply chain resilience: definition, review and theoretical foundations for further study. Int J Prod Res 53(18):5592–5623
Tuncel G, Alpan G (2010) Risk assessment and management for supply chain networks: a case study. Comput Ind 61(3):250–259. https://doi.org/10.1016/j.compind.2009.09.008
Van Hoek RI, Weken HAM (1998) The impact of modular production on the dynamics of supply chains. Int J Logist Manag 9(2):35–50
Ventola CL (2011) The drug shortage crisis in the united states: causes, impact, and management strategies. Pharm Ther 36(11):740–742, 749–757
Visich JK, Li S, Khumawala BM, Reyes PM (2009) Empirical evidence of RFID impacts on supply chain performance. Int J Oper Prod Manag 29(12):1290–1315
Watson JT, Gayer M, Connolly MA (2007) Epidemics after natural disasters. Emerg Infect Dis 13(1):1–5
Wahyu C, Wilujeng S, Chrismardani Y (2022) Uncertain supply chain. Management 10:471–476. https://doi.org/10.5267/j.uscm.2021.12.004
Womack JP, Jones DT (2005) Lean solutions: how companies and customers can create wealth together. Simon & Schuster, New York, NY

World Health Organization (1994) Lexicon of alcohol and drug terms. World Health Organization
Yaroson EV, Breen L, Hou J, Sowter J (2021) Advancing the understanding of pharmaceutical supply chain resilience using complex adaptive system (CAS) theory. Supply Chain Manag Int J

Improving Medical Supply Chain Disruption Management with the Blockchain Technology

Özden Özcan-Top

1 Introduction

Supply chains (SCs) in all domains are subject to disruptions due to natural, human-related, system-related, and financial factors (Ambulkar et al. 2015, Messina et al. 2020; Ali et al. 2021). These factors, also called disruption drivers, mainly include earthquakes, diseases, floods, technology failures, plant fires, political instabilities, and bankruptcies. (Kleindorfer and Saad 2005; Xiao and Yu 2006; Trkman and McCormack 2009; Ali et al. 2021; Singh et al. 2021). Supply chain disruptions refer to the catastrophic events experienced at different components of a supply chain (Tang 2006), such as partial or complete closure of manufacturing plants, interruptions in transportation, shortages of raw materials and labor, and export/import challenges (Tummala and Schoenherr 2011). Such disruptions may result in financial losses, decreased sales, and loss of lives or injuries, depending on the type of industry (Dolgui and Ivanov 2021).

Compared to predictable SC disruption risks, the impact of unforeseen events would be higher, especially in unprepared systems (Chowdhury et al. 2021). *Resiliency* is defined as "the ability of a system to return to its original state or move to a new, more desirable state after being disturbed" (Christopher and Peck 2004). In case of unforeseen events, the resiliency level of the supply chain systems becomes the determining factor in recovery and reducing loss or damage from disruptions.

For the last three years, we have been fighting the coronavirus pandemic (Covid-19), which has seriously affected human lives, the environment, businesses, and national economies (Hobbs 2020; Nasir et al. 2022). Among many other domains, the medical supply chain may be the one that received the hardest hit by the pandemic

Ö. Özcan-Top (✉)
Graduate School of Informatics, Middle East Technical University, Ankara, Türkiye 06800
e-mail: ozdenoz@metu.edu.tr

S. K. Paul et al. (eds.), *Supply Chain Risk and Disruption Management*, Flexible Systems Management, https://doi.org/10.1007/978-981-99-2629-9_11

(Singhet al. 2021; Singh and Parida 2022). The *medical supply chain* can be defined as the flow of medicines and other medical goods from a manufacturer to a patient (Singh and Parida 2022).

One of the primary purposes of a medical supply chain is to ensure the timely and continuous delivery of medicines and medical devices to patients. It is expected from a medical supply chain to respond to the impact of disruptions effectively. However, especially at the initial stage of the Covid-19 outbreak, several cases indicated that such systems failed to manage disruptions' consequences.

The European Medicines Agency has announced that demands have increased dramatically for anesthetics, cardiac and respiratory medicines, antibiotics, and muscle relaxants, which are essential for patients with Covid-19 (European Medicines Agency 2020). The FDA fact sheet (2021) on mitigating and preventing medical device shortages states that the pandemic hit not only the delivery of medicines and personal protective equipment, but also critical devices such as ventilators and test supplies. The FDA also recognizes that the pandemic has impacted public health and drug development programs, ongoing manufacturing operations, and the FDA's ability to conduct inspections (FDA Guidance for Industry 2021).

On the other hand, although the pandemic had severe adverse effects on medical supply chains, it did not bring new challenges. Besides, it revealed the issues and magnified the impact of the problems in the supply chains (Queiroz et al. 2020). The FDA fact sheet (2021) also highlights that Covid-19 revealed the weaknesses in the U. S. medical supply chain, particularly the dependence on raw materials and medical device components from China and other countries.

Unfortunately, established methods are still ineffective in managing risks, and disruptions emerge from unforeseen events (Remko 2020; Chowdhury et al. 2021). A recent review study states that most SC disruption management studies focus on disruption impact analysis at strategic, tactical, and operational levels (Dolgui and Ivanov 2021). However, the authors emphasize the importance of innovative information technologies for developing resilient supply chain systems and mitigating ripple effects (Remko 2020; Dolgui and Ivanov 2021).

Being resilient to disruptions require getting ahead of the problems and being proactive. Receiving accurate and timely information is highly important in dealing with SC disruptions. Following the rapid outbreak of the Covid-19 pandemic, the FDA tried reaching out to more than 1,000 medical device manufacturers to request supply chain information; however, only about one-third responded, often with missing information (The FDA Fact Sheet 2021). Such vulnerabilities exposed in the pandemic in medical SC systems in the United States led to the establishment of the Resilient Supply Chain and Shortages Prevention Program by the FDA with a $21.6 million budget (FDA CDRH Annual Report 2021).

This huge investment made for improving resiliency in medical SCs indicates the significance of the problems. Therefore, we need to find ways to apply preventive measures, identify potential medical product supply shortfalls, and continue surveillance and rapid intervention with emerging technologies such as artificial intelligence, big data management, and blockchain (Choi et al. 2019; Govindan et al. 2020; Goodarzian et al. 2021; FDA CDRH Annual Report 2021).

This chapter explores how problems associated with medical SC disruption management could benefit from one of the most secure technologies that exist for now: the blockchain. The *blockchain (BC)* is a distributed ledger with immutability characteristics that provides transparency, accuracy, and security of the data stored in nodes and transactions (Nofer et al. 2017; Kouhizadeh et al. 2021). It is expected that the adoption of blockchain technology in the healthcare industry would save \$100–\$150 billion per year by 2025 due to savings achieved in the prevention of data breaches and reduction of counterfeit drugs, IT, and operations costs (Arsene 2022). When drug and medical device shortages occur due to disruptions, medical SC stakeholders affected by the shortages need to inform the legal authorities and regulatory bodies. Applications developed based on blockchain technology, such as Medledger (Uddin 2021) and originChain (Lu and Xu 2017), enable collaboration among SC stakeholders and ensure transparency, accuracy, and information flow security.

Although BC technology cannot be a silver bullet in solving all challenges regarding SC disruption management, the area is full of potential. This chapter contributes to the medical SC disruption management literature with a comprehensive overview of BC technology applications in medical supply chain systems following the Covid-19 pandemic. It responds to the following research questions based on the evidence from the literature:

RQ1: What are the impacts of disruptions in a medical supply chain?
RQ2: How could blockchain technology improve the resiliency of medical supply chain systems in dealing with disruption-related risks?

The disruption risks and associated blockchain-based solutions presented in this chapter intend to create awareness of the benefits of using blockchain technology applications in medical supply chains. Furthermore, it provides a starting for practitioners and managers to adopt these technologies in their systems and take the right actions while expecting unforeseen events to occur.

The rest of the chapter is organized as follows: Background information on blockchain technology is given in Sect. 2. In Sect. 3, the blockchain-integrated medical supply chain structure is described. In Sect. 4, medical supply chain disruption risks are presented in five categories, along with the mitigation strategies to solve these issues using blockchain technology. In Sect. 5, the managerial implications of the study are discussed. Finally, the conclusions are presented in Sect. 6.

2 Background on the Blockchain Technology

Blockchain networks are trusted data management systems enabling data transparency and transaction traceability to users. They are developed in a distributed manner without a central authority and provide a secure environment resilient to data manipulations (Yaga et al. 2018). Blockchain technology provides a resilient system for data management. Once data is added to a blockchain network, it cannot be deleted or modified without the approval of network participants (Yaga et al. 2018).

The roots of blockchain technology go back to the early 1990s. Several researchers have contributed to the development of varying components of the technology. Lamport (1998) redesigned distributed systems using a new state machine approach to ensure agreement on unreliable networks. Haber and Stornetta (1990) developed a method for securing time stamps of digital documents so that no one could change them. Nakamoto (2008) brought these components together and published the Bitcoin electronic cash system.

Immutability in blockchains is achieved by calculating the hash value of data and replicating this hash value in every node (Jamil et al. 2019).

Blockchain can be categorized as public and private (Yaga et al. 2018). Public blockchains are entirely decentralized, allowing anyone to join the network. Every node in a public blockchain has equal rights in creating new nodes and validating existing nodes. However, private blockchains are maintained by a central authority. In addition to public and private types, blockchains could be categorized as permissionless and permissioned, where participants do not need to be approved or need to be approved before joining a network, respectively (Yaga et al. 2018).

Smart contracts are another property of blockchain technology, which can be considered as the digital counterparts of real-world contracts (Macrinici et al. 2018). The smart contract concept, which integrates secure computer protocols with user interfaces in public networks, was first introduced by Nick Szabo (1997). They refer to digital agreements between parties that sign them and execute themselves when predetermined conditions are met (Law 2017).

Although blockchain technology has become popular with the development of cryptocurrencies, it has several application areas, from music royalty management to art dealing, digital voting to supply chain management (Baysal et al. 2021).

Figure 1 below presents the structure and working principles of blockchain. In this representation, node A wants to send data to node B. In (1) A block is created in the blockchain containing the transaction information. Every block contains the number and the value of the previous blocks. In (2) the blocks are sent to all network participants. (3) If there is no error or manipulation in the data, participants record the block in their database copy. (4) Afterward, the initial block can be added to the blockchain with all previous transactions' information. Finally, data reaches node B (Sinenko and Doroshin 2020).

3 Blockchain Integrated Medical Supply Chain

Medical supply chain stakeholders include suppliers, manufacturers, warehouses, hospitals/other healthcare centers, regulatory agencies, insurance companies, governmental agencies, pharmacies, healthcare professionals, and patients from a broad perspective. The following figure presents a blockchain-integrated medical supply chain application adopted by Khezr et al. (2019) (Fig. 2).

In Figure 2, In Step 1, a block with a unique ID is created in a digital ledger after the production of each medicine. The transaction includes applied procedures, patent

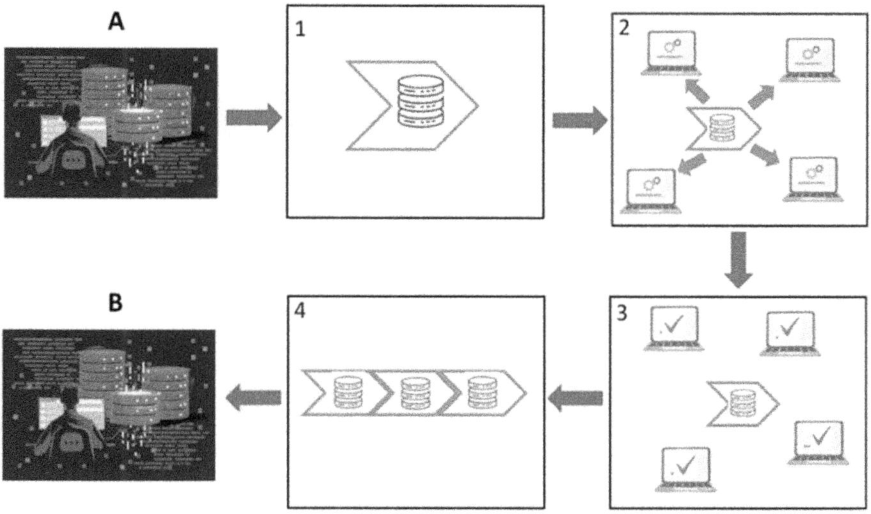

Fig. 1 Work principles of blockchain. Adapted from Sinenko and Doroshin (2020)

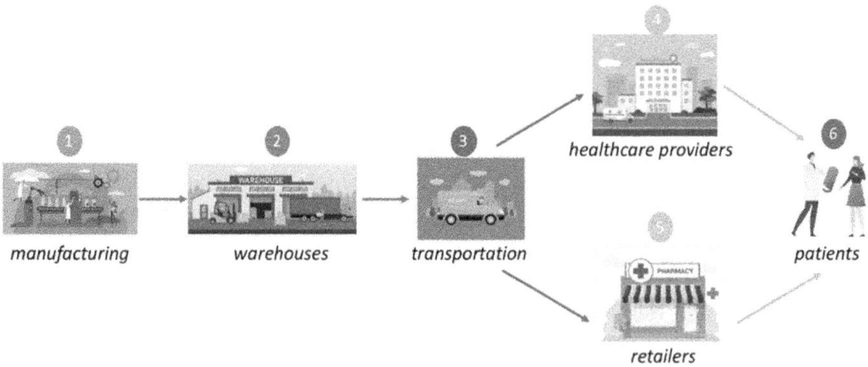

Fig. 2 Medical supply chain stakeholders. Adapted from Khezr et al. (2019)

information, expiry date, and other relevant information. In Step 2, medicines are transferred to warehouses with attached information such as time, lot number, and warehouse identification number. In Step 3, medicines are transferred to healthcare providers, retailers, and pharmacies. Transportation transaction includes time-out and time-in data, the type of transportation in the blockchain. A new transaction is added to the blockchain from one distributor to another in each transportation. Medicines are delivered to patients by healthcare providers and pharmacies by associating patients with the medicines (Khezr et al. 2019).

As every transaction and new information is stored in the blockchain in an immutable way, the whole process can be traced and verified.

4 Medical Supply Chain Disruptions, Associated Risks, and Improving Resiliency via Blockchain Applications

In the course of catastrophic events such as the Covid-19 pandemic, we expect chaos to increase and more harmful situations to be experienced than usual. When a disruption is experienced in the supply chain, whether due to unforeseen or predictable events, its effects are experienced at the physical/operational, information, and financial levels (Blackhurst et al. 2005; Sawik 2013) and discovery, recovery, and redesign processes are applied (Macdonald and Corsi 2013; Messina et al. 2020).

This section focuses on the potential risks that the medical supply chain stakeholders may experience due to disruptions and the possible blockchain-based solutions for each risk item.

In Table 1, supply chain disruptions that may occur in case of unforeseen events and associated risks that could be mitigated using blockchain technologies are listed.

Risk 1: Production of fake (counterfeit, falsified) and substandard medicines: Production of fraudulent medicines increases when shortages are experienced in producing or delivering medicines due to unforeseen events (WHO Drug Information 2006; Tseng et al. 2018). Fake drugs which are not properly formulated or contain dangerous ingredients remain one of the significant issues in the medical industry (Tadeg and Berhane 2012; Sylim et al. 2018; Jamil et al. 2019). They are associated with increased mortality and morbidity and threaten public health and the economy (Kelesidis and Falagas 2015). Even in usual times, it is estimated that 10–30% of the medicines in developing countries are fake (Health Research Funding 2017). The OECD Task Force (2020) report states that counterfeit drug problems have increased with the Covid-19 pandemic due to disrupted supply chains and increased demand for medicines, virus tests, and limited law enforcement capacities. In fragile times such as dealing with a pandemic, the production of substandard medicines that do not follow specific production regulations also increases.

Table 1 Disruptions and associated risks on medical supply chain

SC disruptions	Associated risks
Production Interruption and Delays in delivering medical products	**Risk1**: Production and delivery of fake (counterfeit, falsified) and substandard medicines
	Risk 2: Delivery of medicines by unauthorized parties
	Risk 3: Breakage of cold chains
Interruption in the SC information flow	**Risk 4**: Receiving inaccurate information from various medical SC stakeholders
Interruption in the SC network security	**Risk 5**: Leakage of sensitive patient data, data manipulation, and data theft

Several techniques were developed to determine falsified and counterfeit medicines, such as RFID and barcode scanning in the previous years. However, these technologies are prone to security attacks, such as man-in-the-middle (Kumar and Tripathi 2019).

Mitigation Strategy for Risk 1: Tracking the whole medical SC flow and detecting fake (counterfeit, falsified) and substandard medicines in a medical supply chain using blockchain-based applications: Ensuring security is essential for reliable and resilient supply chain systems. Blockchain technology provides a secure medium to store drug information and detects data anomalies and unauthorized data entries in medical SC networks (Sylim et al. 2018; Haq and Esuka 2018). The path (all transactions and approvals) that medicines take from raw material supply to their delivery to a patient can be monitored in a medical SC. Smart contracts and key verifications are the components of blockchain-based applications that enable transparent information tracking (Kumar and Tripathi 2019; Uddin 2021). Each data entry in a block is time-stamped and is stored by every participant in a network. This way, any attempts to change accurate data could be noticed easily.

Researchers state that blockchain may be the most secure data storage system as once data is added to a blockchain, it cannot be deleted or modified (Haq and Esuka 2018). Several blockchain-based applications were developed to secure medical SC. For instance, The ***Gcoin*** and the ***Medledger*** applications developed by Tseng et al. (2018) and Uddin (2021) ensure transparency in transactions and prevent counterfeit medicines by bringing manufacturers, government agencies, wholesalers, hospitals, and pharmacies together in a transparent, traceable, and verified chain.

Risk 2: Delivery of medicines by unauthorized parties: To run a high-quality supply chain management system with all aspects, we need a balance between time, effort, and budget to perform all SC activities. In catastrophic events, when there is a high demand for certain products and time pressure, this balance among SC quality triangle components is broken. Manufacturers, healthcare providers, and other SC stakeholders need to make quick decisions to specify new partners contributing to an SC. Due to insufficient verification and review processes and workforce, unauthorized parties may leak into medical supply chains leading to corruption in varying components of an SC (Baysal et al. 2021).

Mitigation Strategy for Risk 2: Using blockchain to prevent delivery of medicines by unauthorized parties: Blockchain-based supply chain applications enable monitor of the whole transactions by regulatory bodies. In a chain, observatory accounts can be created, and whenever a transaction occurs, these bodies (e.g., the FDA) are notified. In this way, it is possible to verify the transactions and the parties involved in these transactions (Sylim et al. 2018). If the BC-based application is established based on the "permissioned" rule, only authorized and trusted participants are allowed to perform a transaction in the network (Haq and Esuka 2018).

IBM has also been working on this issue and recently launched a blockchain network called ***Rapid Supplier Connect***. The application enables government agencies and healthcare organizations to quickly identify alternative suppliers and equipment vendors where shortages are experienced in a medical SC (Landi 2020).

Risk 3: Breakage of cold chain: A critical aspect of medical supply chains is to ensure environmental conditions such as temperature, light, ventilation, sanitation, and humidity at certain levels during the logistics of medicines (Bogataj et al. 2005; Bamakan et al. 2021). Constant temperature and humidity levels must be ensured for sensitive medicines in the whole lifecycle from production to final delivery. The systematic review study of Hanson et al. (2017) on evaluating vaccine exposure temperatures states that among the reviewed 45 articles, "*the vaccine exposure temperatures were below the recommended ranges during shipments in 38% of the studies in higher-income countries and 19.3% in lower-income countries*".

The Guidelines on Good Distribution Practices of the World Health Organization (QAS/04.068) and European Union (94/C63/03) state that recorded temperature monitoring data should be available for review, and this data should be available "*at least the shelf life of the stored product plus one year*." In the course of catastrophic events, the manual evaluation of environmental conditions may be interrupted. The monitoring process also requires the involvement of many people and lots of paperwork (Kshetri 2018). These necessities have to be avoided in the circumstances such as a pandemic. Although IoT (Internet of Things), sensor, and RFID (radio-frequency identification) technologies have significantly improved the monitoring of environmental conditions, these devices are still vulnerable to hacking and alterations (Kshetri 2018).

Mitigation Strategy for Risk 3: Using smart contracts to monitor cold chains: Smart contract-integrated IoT devices enable traceability and tracking of the environmental conditions of products in cold chains (Bocek et al. 2017). This way, cold chain stakeholders can be permitted to monitor products' environmental conditions in real time (Bamakan et al. 2021). None of the involved parties can modify the data; thus, trust is ensured among SC stakeholders. Monitoring temperatures and other environmental conditions prevent waste, protects the data from being stolen, and manipulated (Bamakan et al. 2021).

Kshetri (2018) presents the Modum[1] case as an example of using smart contracts to monitor environmental conditions in medical SCs. Modum's system focuses on monitoring temperature levels of ambient medicines, which can be stored at 15–25 °C in trucks. The measured temperature values are transferred to the blockchain when medicines reach a destination. Later on, a smart contract compares all the data against regulatory requirements. If the environmental conditions are met, the medicines are deployed. If not, regulatory bodies are informed about the deviations automatically.

[1] https://www.modum.io.

Risk 4: Leakage of sensitive patient data, data manipulation, and data theft: In unforeseen catastrophic events, medical supply chains become more vulnerable to data manipulation and theft (Kshetri 2018; Baysal et al. 2021). The evidence from the field reveals that patients may be exposed to economic and social threats due to a lack of adequate security measures and data breaches (Kemkarl and Dahikar 2012; Yue et al. 2016).

Mitigation Strategy for Risk 4: Ensuring accurate information flow in a medical SC and preventing data manipulation and theft: To improve medical supply chain resilience and decision-making, we need to ensure the accurate and secure information flow from all SC parties. A blockchain network has the capability to present existing inventory in warehouses, hospitals, and pharmacies, which would enable the rerouting of medicines and other materials where they are most needed (Landi 2020). Different parties in a medical SC may keep varying data regarding patients.

For better decision-making, patient data, including diagnosis, genetic, and biometric information, must be consolidated across different healthcare institutions (Hölbl et al. 2018). Especially in infectious diseases, a large amount of data from an extensive network need to be gathered. Blockchain-based applications ensure that such sensitive information is securely shared and validated by authorities (Messina et al. 2020).

Risk 5: Receiving inaccurate information from various medical SC stakeholders: Shortage of raw materials and increased demand for medical products cause difficulties in delivering medicines to patients. When information flow is corrupted, accurate demand for medicines, tests, and other materials may not be specified. In such cases, optimal use of existing drugs and healthcare equipment becomes more important. Corruptions in the information flow indicating the stock levels and demand would be misleading decision-makers in SCs (Govindan et al. 2020).

Mitigation Strategy for Risk 5: Flexibility in reflecting changes to a medical supply chain by using blockchain: In unexpected events such as the COVID pandemic, regulations are needed to be updated promptly by regulatory agencies. Supply chain systems need to be flexible to reflect such changes in every SC component. The ***originChain*** application developed by Lu and Xu is an example of such a need. It ensures medical data's availability to service providers and automates regulatory-compliance checking in a medicine supply chain (Lu and Xu 2017).

5 Managerial Implications of the Study

There is a great potential for using emerging technologies in the flow of medicines and medical devices, from raw material acquisition to final delivery to the patients. However, as Liza et al. (2022) stated, most companies do not have the awareness or capability to use these technologies.

The disruptions experienced in the Covid-19 pandemic revealed the weaknesses and vulnerabilities of medical SCs and amplified their impacts.

From the management perspective, this experience could be used as an opportunity to make the supply chains better prepared for the consequences of unforeseen events, preventing disruptions and efficiently handling disruptions' impacts.

To minimize risks in medical SCs that emerge from disruptions, we can design and develop holistic solutions that bring together all supply chain components and recent advancements in information technologies.

This study presents disruption-related risks that could be mitigated or resolved using blockchain applications and highlights real-life solution examples. Understanding these risks and potential BC-based solutions would help SC managers and practitioners specify threats and develop recovery plans and actionable strategies to become more agile in times of crisis.

Blockchain has the potential to improve the visibility of medical SCs, stating the sources of the raw materials, manufacturers, warehouses, transportation units, environmental conditions, and final consumers in an immutable way. It should be noted that disruption-related risks can be prevented, and disruptions' short and long-term impacts on medical SCs can be minimized.

Additionally, although this study explores the possible utilization of blockchain technology in medical SC systems, the solutions discussed can be adapted to other SCs.

6 Conclusions

Unforeseen events such as disease outbreaks, earthquakes, or floods are realities of our world, and it is impossible to avoid most of them. However, the impact of disruptions caused by these events on SCs can be reduced, recovery plans can be put into operation effectively, and the resiliency of SCs can be improved with the integration of emerging technologies.

The Covid-19 pandemic has revealed the vulnerabilities and weaknesses of medical supply chains. Production of medicines and other medical products was interrupted, and delays were experienced in delivering products to patients. Decreases in production capacities, shortages in raw materials, and breaks in transportation initiated other problems such as the production and delivery of counterfeit and substandard medicines, involvement of unauthorized parties to SCs, breakage of cold chains, data manipulation, and data theft.

This chapter focused on the risks that emerge from medical supply chain disruptions and discusses how blockchain could be a solution for overcoming these risks and creating more resilient medical supply chains.

The evidence from the literature shows that blockchain has been adopted in several supply chain applications in the medical domain. We can use blockchain technology in developing applications to ensure full traceability of an SC from raw material supply to final delivery, detect fake (counterfeit, falsified) and substandard medicines, monitor environmental conditions, and ensure that cold chains are not broken during the delivery of sensitive medicines, prevent data manipulation and theft, and increase the efficiency of regulatory audits when the SC resources are substantially low due to catastrophic of the events.

By benefiting from the immutable and secure nature of blockchains and ensuring security and privacy, it would be possible to mitigate and proactively reduce the impact of risks that occur due to disruptions. In the future, we expect regulatory bodies and SC stakeholders to increase their investments in adopting blockchain technology in supply chain management.

This study presents the risks and blockchain-based mitigation strategies based on a literature review and has limitations. First, performing a multivocal review and including grey literature (e.g., blog posts, white papers, webinars) in the analysis would extend the number and variety of blockchain-based applications used in the field. Secondly, the review focused solely on the medical supply chain domain. Other SC domains, such as food and automotive, would have commonalities in types of risks and challenges. There may be varying BC applications used in other domains. Therefore, the review could be extended to include other SC domains.

The pandemic is not over yet, and uncertainties remain valid in many ways. The adoption of blockchain by medical supply chain stakeholders needs to be considered as part of their strategic plans to overcome the challenges of unforeseen events.

References

Ali SM, Paul SK, Chowdhury P, Agarwal R, Fathollahi-Fard AM, Jabbour CJC, Luthra S (2021) Modelling of supply chain disruption analytics using an integrated approach: an emerging economy example. Expert Syst Appl 173

Ambulkar S, Blackhurst J, Grawe S (2015) Firm's resilience to supply chain disruptions: scale development and empirical examination. J Oper Manag 33:111–122

Arsene C (2022) The global "Blockchain in Healthcare" report: the 2022 ultimate guide for every executive, healthcare weekly. https://ieeexplore.ieee.org/stamp/stamp.jsp?arnumber=9316803. Accessed 13 May 2022

Bamakan SMH, Moghaddam SG, Manshadi SD (2021) Blockchain-enabled pharmaceutical cold chain: applications, key challenges, and future trends. J Clean Prod 302:127021

Baysal MV, Özcan-Top Ö, Betin-Can A (2021) Implications of blockchain technology in the health domain. Advances in software engineering, education, and e-learning. Springer, Cham, pp 641–656

Blackhurst J, Craighead CW, Elkins D, Handfield RB (2005) An empirically derived agenda of critical research issues for managing supply-chain disruptions. Int J Prod Res 43(19):4067–4081

Bocek T, Rodrigues BB, Strasser T, Stiller B (2017) Blockchains everywhere-a use-case of blockchains in the pharma supply-chain. In: 2017 IFIP/IEEE symposium on integrated network and service management, pp 772–777

Bogataj M, Bogataj L, Vodopivec R (2005) Stability of perishable goods in cold logistic chains. Int J Prod Econ 93:345–356

Choi TM, Wen X, Sun X, Chung SH (2019) The mean-variance approach for global supply chain risk analysis with air logistics in the blockchain technology era. Transp Res Part E: Logist Transp Rev 127:178–191

Chowdhury P, Paul SK, Kaisar S, Moktadir MA (2021) COVID-19 pandemic related supply chain studies: a systematic review. Transp Res Part E: Logist Transp Rev 102271
Christopher M, Peck H (2004) Building the resilient supply chain. Int J Logist Manag 15(2):1–13
Combating Counterfeit Medicine (2006) WHO drug information: geneva. Switzerland 20:3–4
Dolgui A, Ivanov D (2021) Ripple effect and supply chain disruption management: new trends and research directions. Int J Prod Res 59(1):102–109
European Medicines Agency Official Web Site (2020). https://tinyurl.com/bdf6abvv. Accessed 15 May 2022
FDA Fact Sheet, Mitigating and Preventing Medical Device Shortages and Prioritizing Public Health, U.S. Food and Drug Administration. https://www.fda.gov/media/156980/download. Accessed 08 May 2022
FDA Guidance for Industry (2021) Manufacturing, supply chain, and drug and biological product inspections during COVID-19 public health emergency questions and answers. https://www.fda.gov/media/141312/download. Accessed 08 May 2022
FDA Center for Devices and Radiological Health (CDRH), Programmatic Accomplishments, 2021 Annual Report. https://www.fda.gov/media/155793/download#page=5. Accessed 13 May 2022
Funding, Health Research (2017) Shocking counterfeit drugs statistics. https://healthresearchfunding.org/20-shocking-counterfeit-drugs-statistics/. Accessed 04 Apr 2022
Goodarzian F, Taleizadeh AA, Ghasemi P, Abraham A (2021) An integrated sustainable medical supply chain network during COVID-19. Eng Appl Artif Intell 100:104188
Govindan K, Mina H, Alavi B (2020) A decision support system for demand management in healthcare supply chains considering the epidemic outbreaks: a case study of coronavirus disease 2019 (COVID-19). Transp Res Part E: Logist Transp Rev 138:101967
Haber S, Stornetta WS (1990) How to time-stamp a digital document. Conference on the theory and application of cryptography. Springer, Berlin, Heidelberg, pp 437–455
Hanson CM, George AM, Sawadogo A, Schreiber B (2017) Is freezing in the vaccine cold chain an ongoing issue? A literature review. Vaccine 35(17):2127–2133
Haq I, Esuka OM (2018) Blockchain technology in pharmaceutical industry to prevent counterfeit drugs. Int J Comput Appl 180(25):8–12
Hobbs JE (2020) Food supply chains during the COVID-19 pandemic. Can J Agricul Econ/Revue Can D'agroeconomie 68(2):171–176
Hölbl M, Kompara M, Kamišalić A, Nemec Zlatolas L (2018) A systematic review of the use of blockchain in healthcare. Symmetry 10(10):470
Jamil F, Hang L, Kim K, Kim D (2019) A novel medical blockchain model for drug supply chain integrity management in a smart hospital. Electronics 8(5):505
Kelesidis T, Falagas ME (2015) Substandard/counterfeit antimicrobial drugs. Clin Microbiol Rev 28(2):443–464
Kemkarl OS, Dahikar DPB (2012) Can electronic medical record systems transform health care? potential health benefits, savings, and cost using latest advancements in ict for better interactive healthcare learning. Int J Comput Sci Commun Netw 2(3/6):453–455
Kleindorfer PR, Saad GH (2005) Managing disruption risks in supply chains. Prod Oper Manag 14(1):53–68
Kouhizadeh M, Saberi S, Sarkis J (2021) Blockchain technology and the sustainable supply chain: theoretically exploring adoption barriers. Int J Prod Econ 231:107831
Kshetri N (2018) Blockchain's roles in meeting key supply chain management objectives. Int J Inf Manag 39:80–89
Khezr S, Moniruzzaman M, Yassine A, Benlamri R (2019) Blockchain technology in healthcare: a comprehensive review and directions for future research. Appl Sci 9(9):1736
Kumar R, Tripathi R (2019) Traceability of counterfeit medicine supply chain through blockchain. In: 2019 11th ınternational conference on communication systems & networks. IEEE, pp 568–570

Landi H (2020) IBM rolls out blockchain network to address supply-chain issues caused by COVID-19. https://www.fiercehealthcare.com/tech/ibm-rolls-out-blockchain-network-to-match-healthcare-organizations-non-traditional-suppliers. Accessed 05 Apr 2022

Lamport L (1998) The part-time parliament. ACM Trans Comput Syst 16(2):133–169

Law A (2017) Smart contracts and their application in supply chain management. Doctoral dissertation, Massachusetts Institute of Technology

Liza SA, Chowdhury NR, Paul SK, Morshed M, Morshed SM, Bhuiyan MT, Rahim MA (2022) Barriers to achieving sustainability in pharmaceutical supply chains in the post-COVID-19 era. Int J Emerg Mark. https://doi.org/10.1108/IJOEM-11-2021-1680

Lu Q, Xu X (2017) Adaptable blockchain-based systems: a case study for product traceability. IEEE Softw 34(6):21–27

Macdonald JR, Corsi TM (2013) Supply chain disruption management: severe events, recovery, and performance. J Bus Logist 34(4):270–288

Macrinici D, Cartofeanu C, Gao S (2018) Smart contract applications within blockchain technology: a systematic mapping study. Telemat Inform 35(8):2337–2354

Messina D, Barros AC, Soares AL, Matopoulos A (2020) An information management approach for supply chain disruption recovery. Int J Logist Manag 31:489–519

Nakamoto S (2008) Bitcoin: a peer-to-peer electronic cash system. Decent Bus Rev 21260

Nasir SB, Ahmed T, Karmaker CL, Ali SM, Paul SK, Majumdar A (2022) Supply chain viability in the context of COVID-19 pandemic in small and medium-sized enterprises: impli-cations for sustainable development goals. J Enterp Inf Manag 35(1):100–124

Nofer M, Gomber P, Hinz O, Schiereck D (2017) Blockchain. Bus Inf Syst Eng 59(3):183–187

Queiroz MM, Ivanov D, Dolgui A, Fosso Wamba S (2020) Impacts of epidemic outbreaks on supply chains: mapping a research agenda amid the COVID-19 pandemic through a structured literature review. Ann Oper Res 1–38

Remko VH (2020) Research opportunities for a more resilient post-COVID-19 supply chain–closing the gap between research findings and industry practice. Int J Oper Prod Manag 40(4):341–355

Sawik T (2013) Selection of resilient supply portfolio under disruption risks. Omega 41(2):259–269

Singh S, Kumar R, Panchal R, Tiwari MK (2021) Impact of COVID-19 on logistics systems and disruptions in food supply chain. Int J Prod Res 59(7):1993–2008

Singh A, Parida R (2022) Decision-making models for healthcare supply chain disruptions: review and ınsights for post-pandemic era. Int J Glob Bus Compet 1–12

Sinenko SA, Doroshin IN (2020) Use of modern means and methods in the organization and management in construction. IOP Conf Ser: Mater Sci Eng 753(4):042017 (IOP Publishing)

Sylim P, Liu F, Marcelo A, Fontelo P (2018) Blockchain technology for detecting falsified and substandard drugs in distribution: pharmaceutical supply chain intervention. JMIR Res Protoc 7(9):e10163

Szabo N (1997) Smart contracts: formalizing and securing relationships on public networks. First Monday 2(9). https://doi.org/10.5210/fm.v2i9.548

Tadeg H, Berhane Y (2012) Substandard and counterfeit antimicrobials: recent trends and implications to key public health interventions in developing countries. East Afr J Public Health 9(2):85–89

Tang CS (2006) Perspectives in supply chain risk management. Int J Prod Econ 103(2):451–488

The OECD Task Force (2020) Trade in fake medicines at the time of the covid-19 pandemic. https://www.oecd.org/gov/illicit-trade/oecd-fake-medicines-webinar-june-10-summary-note.pdf. Accessed 07 Apr 2022

Tseng JH, Liao YC, Chong B, Liao SW (2018) Governance on the drug supply chain via gcoin blockchain. Int J Environ Res Public Health 15(6):1055

Trkman P, McCormack K (2009) Supply chain risk in turbulent environments—A conceptual model for managing supply chain network risk. Int J Prod Econ 119(2):247–258

Tummala R, Schoenherr T (2011) Assessing and managing risks using the supply chain risk management process (SCRMP). Supply Chain Manag 16(6):474–483

Uddin M (2021) Blockchain medledger: hyperledger fabric enabled drug traceability system for counterfeit drugs in pharmaceutical industry. Int J Pharm 597:120235

Xiao T, Yu G (2006) Supply chain disruption management and evolutionarily stable strategies of retailers in the quantity-setting duopoly situation with homogeneous goods. Eur J Oper Res 173(2):648–668

Yaga D, Mell P, Roby N, Scarfone K (2018) Blockchain technology overview—National Institute of Standards and Technology Internal Report 8202. https://doi.org/10.6028/NIST.IR.8202. https://csrc.nist.gov/publications/detail/nistir/8202/final. Accessed 08 Apr 2022

Yue X, Wang H, Jin D, Li M, Jiang W (2016) Healthcare data gateways: found healthcare intelligence on blockchain with novel privacy risk control. J Med Syst 40(10):1–8

Impacts of Resilience Practices on Supply Chain Sustainability

Noraida Azura Darom and Hawa Hishamuddin

1 Introduction

Supply chains are continuously exposed to the possibility of disruption as the interconnected nature of their operation creates many vulnerabilities. The interdependence between suppliers, manufacturers, and retailers in the supply chain can lead to various unforeseen situations, subsequently exposing organizations to the risk of disruption. Without an effective strategy, the uncertainty and unpredictable events that could occur might lead to poor supply chain performance and severe loss. Furthermore, major disruptions could cause lasting effects throughout the supply chain. Thus, the importance of supply chain disruption management has been widely emphasized and has become even more crucial in the current global scenario.

The recent virus Covid-19 outbreak has shown how various industries were affected by the major disruption. The impact of the pandemic spans the automotive sector, tourism industry, aviation industry, oil industry, construction industry, telecommunication sector, food industry, and healthcare industry. Supply and sourcing strategies were disrupted because of delayed or stopped manufacturing processes due to the lack of stock, manpower, and travel restrictions. While industries like tourism or aviation suffered from lockdown and travel restrictions, the healthcare supply chain on the other hand was faced with increased demand as the shortage of supply of personal protective equipment and medical equipment affected all around the world.

In India, Singh et al. (2021) reported that the food industry supply chain has been severely affected as the lockdown orders caused the unavailability of the labor force

N. A. Darom (✉) · H. Hishamuddin
Department of Mechanical and Manufacturing Engineering, Faculty of Engineering and Built Environment, Universiti Kebangsaan Malaysia, Bangi, Malaysia
e-mail: noraida.darom@gmail.com

H. Hishamuddin
e-mail: hawa7@ukm.edu.my

S. K. Paul et al. (eds.), *Supply Chain Risk and Disruption Management*, Flexible Systems Management, https://doi.org/10.1007/978-981-99-2629-9_12

and truck operators. Small and medium industries had to suspend manufacturing, while large-scale manufacturing firms had to reduce production capacity to control the outbreak (Udofia et al. 2021). Similarly, in Pakistan, many manufacturing activities were facing substantial delays as their suppliers, which are mainly based in China, are operating at a limited capacity (Butt 2021).

The impact of a pandemic has demonstrated the importance of developing a resilient supply chain for operational continuity. However, alongside the need to develop resiliency in the supply chain, sustainability goals should continue to become the priority for the supply chain. Achieving environmental sustainability in the supply chain would help to reduce costs, improve safety, and protect the environment in addition to increasing customer satisfaction and loyalty. Sustainability measures could range from a small operational improvement, adoption of green technology, or the extent of reconfiguring the whole supply chain, all with the ultimate goal of better environmental performance.

Many researchers have suggested that a supply chain with high resilience would be more sustainable. Mehrjerdi and Shafiee (2021) claim that a supply chain should be resilient enough to maintain its sustainability. However, as the organization takes measures to build resiliency into the supply chain by developing the ability to prepare for, address, and restore operational continuity (Golgeci and Ponomarov 2013), these courses of action could compromise the sustainability aspect of the supply chain. The supply chain practices in managing disruption risks could have a significant impact on the environmental sustainability of the supply chain (Govindan et al. 2014).

Thus, the motivation of this chapter is to address the relationship between strategies employed by a supply chain to build resiliency against disruption and the impact of those practices on the sustainability aspects of the supply chain. Understanding the relationship between resilient practices and sustainability would be important to balance the two supply chain goals.

This chapter could serve as a starting place to gain an understanding of the concept of a resilient and sustainable supply chain, and key strategies to achieve these goals. The chapter is organized as below. In Sect. 2, we present a summary of resilient strategies for dealing with supply chain disruptions. Next, the environmental sustainability aspect of the supply chain is reviewed in Sect. 3. Section 4 discusses the relationship between the resilient practices and sustainability aspect of a supply chain, Sect. 5 presents the literature on the quantitative models that were developed to address the supply chain problem concerning sustainability and resiliency issues. Section 6 presents some observations and insights, while Sect. 7 ends the chapter with the conclusion.

2 Resilient Practices in Supply Chain Disruption Management

Risks in the supply chain can be defined as disruption, uncertainty, disasters, and dangers that are inherent in the supply chain (Ghadge et al. 2013). According to Lynch (2012), any change that occurs in the supply chain can lead to potential risks that interfere with the operation of the entire system. In addition, the source of these failures or disruptions can occur regardless of the scale and is not limited to a function, process, source, or location. Even without a large-scale unprecedented event, disruptions can appear in the daily operation of a supply chain when there are risks of failures in machine operation, slow deliveries, poor planning and scheduling, lack of raw materials, as well as human or employee fault (Kilpatrick 2020). Thus, the concept of a resilient supply chain is based on the rationale that not all risks are avoidable but by building resilience, an organization can address the threat of disruption to their supply chain and ensure continued delivery of products and services to customers (Tukamuhabwa et al. 2017).

Literature offers different classifications of the type of risk. Chopra and Sodhi (2004) categorizes supply chain risk into demand risk, intellectual property risk, risk behavior, and political risk or social. Ghadge et al. (2013), classify the types of risks into six key characteristics: process, organization, location, data, application, as well as technology. Process risk revolves around the sequence of activities in the organization and their end results while organizational risk focuses on human resources concerning their capabilities and roles as well as the relevant team structure and organizational units. Meanwhile, location risk is associated with geographical location and related issues with physical facilities and infrastructure, while data risk focuses on means of addressing the content, structure, and relationships associated with information data. In addition, application risk refers to the capabilities and usability of information systems. Alongside that, technological risk involves the equipment and technology application.

The strive to build resiliency in the supply chain would serve as the long-term strategy for an organization to handle unexpected and continuous disruptions. With this capability, the organization could proactively plan the supply chain to predict unexpected disruptions, respond to disruptions continuously, be able to control structure and function, and be able to maintain robust or better operations to gain advantage in competitiveness (Ponis and Koronis 2012). By being resilient, the supply chain would be able to reduce the impact of vulnerability (due to disruption) through the expansion of required levels of readiness, rapid response, and recovery capabilities (Chowdhury and Quaddus 2015). From a much wider perspective, resilience has been defined as the adaptive capacity where the system is capable to adapt to new conditions and improve other dimensions of performance (Burnard et al. 2018).

Literature has discussed several supply chain capabilities that are strongly linked to resiliency in the supply chain. These include agility, flexibility, responsiveness, adaptability, alignment, robustness, and redundancy (Christopher and Peck 2004; Mensah and Merkuryev 2014; Ponomarov and Holcomb 2009; Sodhi and Tang 2012).

Christopher and Peck (2004) defined several principles for supply chain resilience which include supply chain re-engineering whereby risk reduction is factored in the supply chain design.

According to Shao (2013), responsiveness, competency, flexibility, and speed are characteristics of being agile. Improvement in agility could be achieved by preparing detailed plans and focusing on supply chain policies and initiatives that promote collaboration, integration, and timely information sharing. Through collaboration, faster disruption recovery could be achieved as supply chain parties could coordinate their responses and share resources. Kamalahmadi and Mellat-Parast (2016) suggested that developing contingency plans using flexibility in suppliers' production capacity is an effective strategy for firms to mitigate the severity of disruptions. In addition, the flexibility and reliability of the suppliers and regions play a significant role in determining contingency plans during disruption. Tukamuhabwa et al. (2017), also highlighted the need to increase flexibility and supply chain agility while forming collaborative supply chain relationships. Additionally, the power structure in a supply chain governs the type and configuration of supply chain flexibility (Angkiriwang et al. 2014).

According to Chopra and Sodhi (2014), companies can build resilience by isolating or balancing the supply chain and avoiding too much resource centralization. Mensah and Merkuryev (2014) proposed lean production, the six sigma strategy, and emphasized the strong corporate culture required in a resilient organization. Moreover, Ambulkar et al. (2015) emphasized the need for firms to have a strong awareness of disruption risk and the ability to learn from prior disruptions, the ability to reconfigure resources to manage disruption, and have an infrastructure that is well prepared to handle disruption risks. Dubey et al. (2019) argued that supply chain visibility improves resilience in the supply chain as it facilitates information sharing and data connectivity among supply chain partners as well as building trust when each party shows their commitment to the partnership. Hasani and Khosrojerdi (2016) proposed six resilience strategies including facility dispersion, facility reinforcement, production of semi-manufactured products, multiple sourcing, inventory buffer, and use of a primary and alternative bill of material.

The significance of supplier and inventory management systems has been covered in the literature in relation to supply chain resilience strategies. Kamalahmadi and Parast (2017) evaluated the effectiveness of mitigation strategies using backup providers and emphasized that comprehensive supply risk management strategies should take into account the impact of disruption on interconnected suppliers. Additionally, it is important to use selection criteria that can help reduce disruption and its impact, such as supplier technological capabilities, financial and business stability, as well as process reliability. By creating redundancy through the strategic and selective use of spare capacity and inventory, firms can utilize spare stocks, multiple suppliers, and extra facilities to cope with disruptions (Tukamuhabwa et al. 2015). Moreover, inventory management needs to be aligned for the entire supply chain to minimize the risk of inventory shortage or absence. Furthermore, logistics capabilities need to be enhanced by reducing cycle times and delivery times as well as efficient knowledge management and customer service so that rapid disruption recovery can occur (Ambulkar et al. 2015). Razavian et al. (2021) also highlighted the use of multiple suppliers, emergency inventory, and protection of suppliers as essential strategies to build resiliency and minimize costs. Udofia et al. (2021) emphasized productivity management to maintain customer satisfaction during disruption.

Following the Covid-19 pandemic, Martins et al. (2021) highlighted three critical elements of resilience: the decision-making process, human resources development and knowledge management, and data security. Additionally, the authors highlighted the importance of digital technology to enhance resilience. The effect of severe disruption would necessitate investment in process digitization and technologies to improve the analysis and decision-making process. Meanwhile, Ivanov and Dolgui (2021) proposed a framework of supply chain disruption management comprising data-driven disruption modeling in the supply chain and uncovering the interrelations of risk data, disruption modeling, and performance assessment. The digital supply chain model would allow end-to-end visibility to improve resilience and allow contingency plan assessment. The authors emphasized the importance of supply chain monitoring and visibility in post-pandemic recoveries.

3 Environmental Sustainability of a Supply Chain

According to the United Nations (UN), environmental sustainability is about acting in a way that ensures future generations have the natural resources available to live an equal, if not better, way of life than current generations. Under UN Sustainable Development Goal 12 (SDG), among the targets are sustainable use of natural resources, efficient management of chemical and all waste, and a reduction in waste generation by the year 2030. In addition, SDG-13 has been developed specifically to address climate change and its effects (*Responsible Consumption and Production | United Nations ILibrary*). In achieving this ideal, both individuals and institutions must play a role in environmental sustainability. Sustainability is important for organizations not only to achieve compliance with regulations and meet stakeholders' requirements, but also to gain business advantages such as improved business image, productivity, and product quality (Youn et al. 2013; Zailani et al. 2015).

In the supply chain context, the aspect of sustainability is seen as a combination of considerations of economic aspects, environmental impact, and social responsibility with an efficient inter-organizational business system. This includes the management of materials, information, capital flows, as well as procurement processes that are effective and able to meet the needs of stakeholders to increase the profitability, competitiveness, and resilience of the organization for the short and long term (Ahi and Searcy 2013). Environmental impact factors in operations should also be addressed for the long term and on an ongoing basis. Environmental problems originating from numerous industries could include acid rain, air, asbestos, metals, toxic substances, and universal waste (Balon 2020). For example, in the automotive industry, pollution results from inefficient use of resources, and the automotive industry are one of the main pollution sources in Asia (Zailani et al. 2015). Waste materials issue is among the major concern emphasized in the literature including packaging materials waste and scrapped toxic materials (Gupta and Palsule-Desai 2011; Barbosa-Póvoa 2009; Bonney and Jaber 2011). Moreover, traffic congestion has also been highlighted as one of the environmental impacts of supply chain operations (Chin et al. 2015).

Effects of greenhouse gas emissions, especially carbon dioxide emissions from supply chain activities have been heavily mentioned in the literature (Tang and Zhou 2012; Gupta and Palsule-Desai 2011; Eltayeb et al. 2011; Barbosa-Póvoa 2009; Bonney and Jaber 2011). Carbon emissions in the supply chain can be contributed to the procurement process, product manufacturing, distribution and retail, and disposal and recycling. Most of the carbon emissions are accounted for in the transportation and logistics processes of the supply chain (Homayouni et al. 2021). Other mention of environmental concern includes water pollution (Tang and Zhou 2012), soil pollution (Eltayeb et al. 2011), and the production of hazardous material (Gupta and Palsule-Desai 2011) and (Barbosa-Póvoa 2009).

The drive for sustainability would entail the whole parties in the supply chain to be responsible and involved in the initiative. The efforts could be characterized by the approaches taken in managing resource utilization, production processes, and logistics activity that occur daily in supply chain operations in such a way that there will be a minimal harmful impact on the environment. Additionally, the impact on energy consumption is a driver of efforts to increase efficiency in the use of utilities (Tang and Zhou 2012; Chin et al. 2015; Gupta and Palsule-Desai 2011; Eltayeb et al. 2011).

The extent of the implementation of environmental sustainability strategy would differ from one organization to another. A few examples can be offered. Azevedo et al. (2013) presented in their study how one company had changed their plant roof to include transparent roof material so that natural light could be used. The other company in the same study had changed a particular production process to reduce the water and energy consumption at the same time that it reduced the processing time. All the companies employ reverse logistics management, mainly by using returnable/reusable packages and racks, and return of defective items. According to Govindan et al. (2020), sustainability practices either through a green supply chain approach or a sustainable supply chain management can lead to improved operational performance companies through waste reduction, effective use of utilities, employee involvement, as well as community support. Carbon footprint reduction can also be achieved through sustainability program implementation.

4 Relationship Between Resilience Strategies on Supply Chain Sustainability

Environmental practices must be addressed to assure that the management system is sustainable (Carvalho et al. 2011). Sustainable supply chain practices such as green procurement as well as sustainable packaging can have a positive impact on the economic, social, and operational performance of the supply chain (Zailani et al. 2012). According to Azevedo et al. (2013), environmental performance improvement through a green supply chain management approach can reduce the environmental impact of industrial activities without sacrificing quality, cost, reliability, performance, or efficiency of energy use.

There is an argument that sustainability and resilience objectives could be conflicting with each other. Esfahbodi et al. (2016) stated that the adoption of sustainable practices would lead to better environmental performance, but does not necessarily lead to improved cost performance. Similarly, resilient supply chains may not be the lowest cost even though they would possess the capability to cope effectively with disruptions.

In an ideal situation, a supply chain will be resilient and sustainable when it has the required capabilities to respond effectively to disruption while being able to reduce its vulnerabilities (Nayeri et al. 2022). However, according to Mari et al. (2014), it is hard to maintain supply chain network sustainability when the supplier is vulnerable to disruption or when the manufacturer or warehouse is in a risky location.

Fahimnia and Jabbarzadeh (2016) suggested that sustainability practices through the lean and less waste approach could create vulnerability when disruption occurs, as protective redundancy is reduced, thus affecting the disruption management capability. In inventory management, the optimal lot-sizing decision would balance the need for protection against disruption and minimizing carbon emissions. Small and frequent lot-size lead to more carbon emissions due to transportation and increased logistics costs and a greater risk of supply chain disruption, but carbon emissions due to warehouse operations and inventory costs can be saved (Kaur and Singh 2019).

According to Govindan et al. (2014), the use of flexible transportation to mitigate disruption risk will have a significant impact on the environmental sustainability of the supply chain. Furthermore, planning for additional capacity to achieve robustness would require higher supply chain cost (Aldrighetti et al. 2021). Meanwhile, an implementation of a resilient strategy through facility fortification has been reported to increase supply chain resilience and enhances sustainability (Ivanov 2018).

Supplier selection is very important in designing a sustainable supply chain in order to minimize carbon footprint from the purchased materials (Mari et al. 2014). However, sustainable sourcing practices with only a few numbers of suppliers would limit the ability to switch suppliers during disruption.

5 Modeling of Sustainable and Resilient Supply Chain

Even though disruptions could be an everyday occurrence in the supply chain operation, the recent pandemic has amplified the need for effective strategies to mitigate them. As with other optimization models concerning supply chain problems, the modeling approach can be used to assist in the policy and decision-making process. This approach would allow for the exploration of other opportunities and alternative problem solutions. In addition, this also enables a systematic understanding of the whole supply chain system.

The supply chain planning stage is the phase where many significant strategies can be implemented. Thus, the supply chain network design area has been the focus to integrate sustainability into the supply chain model. Similarly, the incorporation of resilient aspects has been approached through supply chain network design (SCND). In the modeling of sustainable and resilient supply chain network design, the goals are set higher wherein the objectives now are to minimize cost and maximize environmental and social performance, while ensuring that the designed supply chain is resilient to supplier disruptions. The expected disruption cost is used to summarize different components that are related to a disruption including damage cost, recovery/restoration cost, trans-shipment cost, procurement cost penalty, backlog cost, delay penalty cost, and transport cost penalty (Aldrighetti et al. 2021).

Shafiee et al. (2021) developed a model integrating leanness, sustainability, and resilience in the supply chain using mathematical modeling. Resilient strategies of multiple sourcing and backup supplier were taken into account in the model. The economic objective function minimizes operational and strategic costs, while the environmental objective function reduces the pollution emitted from installation and transportation.

Salama and McGarvey (2021) proposed a multi-integer linear program model of a resilient supply chain to the pandemic effects. With constraints on workforce limit, operating hours, and shipment size in addition to many limiting factors imposed by the pandemic, different scenarios were studied for the optimal supply chain design and management during the pandemic.

Mehrjerdi and Shafiee (2021) presented a multi-objective mixed-integer programming model for a closed-loop supply chain. In this research, different dimensions of sustainability have been taken into account with consideration for reducing the total cost, energy consumption, pollution, and increasing job opportunities. Information sharing and multiple sourcing strategies have been employed in model development for making the supply chain more resilient.

Mari et al. (2014) proposed a model for a sustainable and resilient supply chain using the weighted goal programming method. The sustainability goal is represented by the carbon footprint of procured materials, and the carbon emission from transportation and manufacturing. The economic goal is to minimize total supply chain network costs. Meanwhile, a resilience metric is used to determine the expected disruption cost or the resilience aspect of the model. The main objective is to ensure that the supply chain network is sustainable, as well as resilient enough to cope with disruption. The case example shows that when the weightage for the total cost goal is high, the sustainability and resiliency of the supply chain would be reduced.

In the SCND model developed by Jabbarzadeh et al. (2016), the authors consider the risk of disruptions, uncertainties in demand, probability of disruption occurrence, and capacity of facilities in the model. The model's objective is to minimize the total cost of establishing the network while maximizing supply chain resilience. The real-world application involved a case study in which the company planned to build fortified storage units to mitigate fire disasters. The finding from the study shows that facility and initial capital investment are vital in developing a resilient supply chain and reducing strategic supply chain costs.

Fahimnia and Jabbarzadeh (2016) presented a model for a sustainable and resilient supply chain where the aim is for sustainability performance to remain unaffected or only slightly affected by disruptions. The sustainability performance scoring approach is used to quantify the environmental and social performance of the supply chain, focusing on suppliers' sustainability performance. In the case study application, dynamic sustainability trade-off analysis is used to design a supply chain that can provide efficient and effective solutions in normal situations and during disruption. Analysis from the study shows that a small increase in the regular cost of the supply chain can lead to the development of a resiliently sustainable supply chain. In another study by Zahiri et al. (2017), a sustainable-resilient supply chain network design is proposed for the pharmaceutical industry to minimize total network cost and environmental impact and maximize social impact. The social measures in this study include the consideration of job creation and unemployment criteria.

Meanwhile, Ivanov (2018) proposed a scheduling model that considers the coordination of recovery actions in the supply chain. A resilience index was proposed by using the notion of attainable sets which are the range of performance outcomes of the control policies in the presence of disruptions. The model could be used to adjust the supply chain production schedule during severe disruption recovery and assess the resilience status of the supply chain.

Jabbarzadeh et al. (2018) proposed a hybrid approach to design a resilient and sustainable supply chain. In the two-step phase, the sustainability assessment of the suppliers was made followed by the development of a stochastic bi-objective optimization model using sustainability scores as the input parameters. The objectives are to determine the sourcing decisions and resilience strategies such as the need for a backup supplier and extra production capacities. In the case problem of a pipe manufacturing company, environmental measures involved safe treatment and disposal of hazardous materials, waste collection, emission of pollutants, and renewable and non-renewable energy consumption. Additionally, the social criteria focused on human rights, labor working conditions, society contributions, and product responsibility issues, while the economic measures included market shares, profitability, and operating expenses. Findings from the study show that the higher the level of sustainability, the higher the total cost of the chain supply would be. Trade-off analysis was performed to identify opportunities to improve performance such that sustainability remains cost-efficient under various disturbance scenarios.

Later, Pavlov et al. (2019) proposed a network redundancy optimization model using the case example of seaport operations. The authors studied a trade-off between sustainable resource utilization and supply chain resilience in the contingency plan for disruption scenarios that are caused by supply and seaport structural disruptions. The proposed model aims to create resiliency strategies related to the design stage, such as backup suppliers, re-routing in transportation channels, and capacity adjustments. Meanwhile, Hosseini et al. (2019) proposed a stochastic bi-objective mixed-integer programming model to support the decision-making on how and when to use both proactive and reactive strategies in supplier selection and order allocation. The supplier resilience cost is considered to include the mitigation cost to reduce vulnerability and the contingency cost to enhance recoverability. Findings from the study showed that total resilience cost will vary according to different disruption scenarios.

Razavian et al. (2021) developed a model for resilient supply chain model with concurrent consideration for material and financial design. Multiple sourcing, emergency inventory, and additional production capacity were considered and the results show that the strategies could improve the supply chain performance. Meanwhile, the study by Homayouni et al. (2021) proposed a sustainable logistics model that considered assorted vehicle types and gas emissions involved with product transportation. Sustainable logistics planning has the objectives to minimize total cost and maximize the vehicle's transportation performance. The authors categorized carbon emissions sources into operation-related and transportation-related while considering two carbon policies: carbon tax and cap-and-trade. The real-world application of the model involves the design of a supply chain for carbon manufacturers with environmental aspect considerations. This study employed a multi-choice goal programming method to solve the problem and the findings suggested that a cap-and-trade policy is a better carbon policy for reducing pollution.

These studies show the different approaches used in the modeling of sustainable and resilient supply chains. Among the resilience strategies considered include expansion of supplier capacity, use of backup suppliers, and resource investment for faster supplier recovery. Total supply chain cost represents the economic aspect of sustainability while the environmental objective is usually represented by the carbon emissions criteria.

6 Observations and Insights

Following a high-impact disruption event like the Covid-19 pandemic, it is expected that there will be much focus and emphasis on building resiliency in the supply chain to ensure operational continuity during challenging times. Elements of resilience and sustainability like flexibility, agility, robustness, green, and collaboration are not new and have been highlighted extensively in the literature. However, the challenges are to understand the interdependence between these elements when combined together. The actual implementation of resilient strategies has been focused on manufacturing flexibility and transportation capability. Collaboration in the supply chain is also crucial for a coordinated response when disruption occurs.

The literature on the resilient supply chain has shown that decisions on inventory management, supplier selection, and facility decision are always important to the supply chain. However, the utilization of digital technology has been highlighted as essential to improve both the resilience and sustainability of the supply chain in the current scenario. The application of digital technology is an important strategy to cope with the disruptions caused by the pandemic as it allows the means to cope with the restriction imposed (Shen and Sun 2021). However, the technology risk would need to be fully understood and an action plan would be needed for possible new risks.

Understanding the relationship between resilient practices and sustainability is important. Literature has shown that different resilient strategies would affect the sustainability of the supply chain differently.

The application of the optimization model can be used to find the trade-off between resilience and sustainability. While resilient supply chains would offer better protection against disruptions, they might be costly to implement and maintain. Sustainable practices, on the other hand, can create environmental benefits that offset some of these costs. There would be a challenge for the supply chain manager to accurately evaluate this cost. By understanding how certain changes to a supply chain might impact both resilience and sustainability, it is possible to make informed decisions about which practices to adopt. Furthermore, successful implementation in real practice would depend on the specific supply chain issue and managerial policy.

7 Conclusions

A supply chain should achieve by some means the balance of economic, social, and environmental components of sustainability. In this chapter, we reviewed the resilient practices employed as mitigation strategies against disruption risks in the supply chain. We discussed the environmental sustainability aspect of the supply chain. In addition, the interrelationship between the resiliency element and the sustainability performance of the supply chain is also examined.

The review of the existing quantitative work on sustainable and resilient supply chains shows how the optimization model is used to meet the different objectives including supplier selection and developing a contingency plan and a logistics plan. Some of the studies highlighted that resilience enhancements can be achieved at only a small cost increase.

Different case study results show how the supply chain planning was designed to the specific objectives of the supply chain. Additionally, the real-world application of the model illustrates the improvement in environmental performance and reduction of vulnerabilities in the supply chain.

The research area of the sustainable and resilient supply chain is still expanding. There is still lacking a comprehensive supply chain solutions to improve resilience and sustainability as supply chain policies and practices should be supported by an effective operational action plan. Further research should consider if a resilient strategy through digital technology application would have an impact on other aspects of sustainability.

Additionally, the decision-making support system using a model-based can be explored from many approaches. The complexities and various supply chain operations offer opportunities for different research perspectives. More research is needed to address and capture the different elements required for a successful sustainable and resilient supply chain implementation.

References

Ahi P, Searcy C (2013) A comparative literature analysis of definitions for green and sustainable supply chain management. J Clean Prod 52:329–341

Aldrighetti R, Battini D, Ivanov D, Zennaro I (2021) Costs of resilience and disruptions in supply chain network design models: a review and future research directions. Int J Prod Econ 235(March)

Ambulkar S, Blackhurst J, Grawe S (2015) Firm's resilience to supply chain disruptions: scale development and empirical examination. J Oper Manag 33–34(1):111–122

Angkiriwang R, Pujawan IN, Santosa B (2014) Managing uncertainty through supply chain flexibility: reactive versus proactive approaches. Prod Manuf Res 2(1):50–70

Azevedo SG, Govindan K, Carvalho H, Cruz-Machado V (2013) Ecosilient index to assess the greenness and resilience of the upstream automotive supply chain. J Clean Prod 56:131–146

Balon V (2020) Green supply chain management: pressures, practices, and performance—An integrative literature review. Bus Strate Dev 3(2):226–244

Barbosa-Póvoa AP (2009) Sustainable supply chains: key challenges. In Comput Aided Chem Eng 27(C) (Elsevier Inc.)

Bonney M, Jaber MY (2011) Environmentally responsible inventory models: non-classical models for a non-classical era. Int J Prod Econ 133(1):43–53

Burnard K, Bhamra R, Tsinopoulos C (2018) Building organizational resilience: four configurations. IEEE Trans Eng Manag 65(3):351–362

Butt AS (2021) Understanding the implications of pandemic outbreaks on supply chains: an exploratory study of the effects caused by the COVID-19 across four South Asian countries and steps taken by firms to address the disruptions. Int J Phys Distrib Logist Manag

Carvalho H, Duarte S, Machado VC (2011) Lean, agile, resilient and green: divergencies and synergies. Int J Lean Six Sigma 2(2):151–179

Chin TA, Tat HH, Sulaiman Z (2015) Green supply chain management, environmental collaboration and sustainability performance. Procedia CIRP 26:695–699

Chopra S, Sodhi MS (2004) Managing risk to avoid supply-chain breakdown. MIT Sloan Manag Rev

Chopra S, Sodhi MS (2014) Reducing the risk of supply chain disruptions. MIT Sloan Manag Rev 55(3):73–80

Chowdhury MMH, Quaddus MA (2015) A multiple objective optimization based QFD approach for efficient resilient strategies to mitigate supply chain vulnerabilities: The case of garment industry of Bangladesh. Omega (United Kingdom) 57:5–21

Christopher M, Peck H (2004) Building the resilient supply chain. Int J Logist Manag 15(2):1–14

Dubey R, Gunasekaran A, Childe SJ, Papadopoulos T, Blome C, Luo Z (2019) Antecedents of resilient supply chains: an empirical study. IEEE Trans Eng Manag 66(1):8–19

Eltayeb TK, Zailani S, Ramayah T (2011) Green supply chain initiatives among certified companies in Malaysia and environmental sustainability: investigating the outcomes. Resour Conserv Recycl 55(5):495–506

Esfahbodi A, Zhang Y, Watson G (2016) Sustainable supply chain management in emerging economies: trade-offs between environmental and cost performance. Int J Prod Econ 181:350–366

Fahimnia B, Jabbarzadeh A (2016) Marrying supply chain sustainability and resilience: a match made in heaven. Transp Res Part E: Logist Transp Rev 91:306–324

Ghadge A, Dani S, Chester M, Kalawsky R (2013) A systems approach for modelling supply chain risks. Supply Chain Manag 18(5):523–538

Golgeci I, Ponomarov SY (2013) Does firm innovativeness enable effective responses to supply chain disruptions? An empirical study. Supply Chain Manag 18(6):604–617

Govindan K, Azevedo SG, Carvalho H, Cruz-Machado V (2014) Impact of supply chain management practices on sustainability. J Clean Prod 85:212–225

Govindan K, Rajeev A, Padhi SS, Pati RK (2020) Supply chain sustainability and performance of firms: a meta-analysis of the literature. Transp Res Part E: Logist Transp Rev 137(March)

Gupta S, Palsule-Desai OD (2011) Sustainable supply chain management: review and research opportunities. IIMB Manag Rev 23(4):234–245

Hasani A, Khosrojerdi A (2016) Robust global supply chain network design under disruption and uncertainty considering resilience strategies: a parallel memetic algorithm for a real-life case study. Transp Res Part E: Logist Transp Rev 87:20–52

Homayouni Z, Pishvaee MS, Jahani H, Ivanov D (2021) A robust-heuristic optimization approach to a green supply chain design with consideration of assorted vehicle types and carbon policies under uncertainty. Ann Oper Res

Hosseini S, Morshedlou N, Ivanov D, Sarder MD, Barker K, Al A (2019) Resilient supplier selection and optimal order allocation under disruption risks. Int J Prod Econ 213(August 2018):124–137

Ivanov D (2018) Revealing interfaces of supply chain resilience and sustainability: a simulation study. Int J Prod Res 56(10):3507–3523

Ivanov D, Dolgui A (2021) A digital supply chain twin for managing the disruption risks and resilience in the era of industry 4.0. Prod Plan Control 32(9):775–788

Jabbarzadeh A, Fahimnia B, Sheu JB, Moghadam HS (2016) Designing a supply chain resilient to major disruptions and supply/demand interruptions. Transp Res Part B: Methodol 94:121–149

Jabbarzadeh A, Fahimnia B, Sabouhi F (2018) Resilient and sustainable supply chain design: sustainability analysis under disruption risks. Int J Prod Res

Kamalahmadi M, Mellat-Parast M (2016) Developing a resilient supply chain through supplier flexibility and reliability assessment. Int J Prod Res 54(1):302–321

Kamalahmadi M, Parast MM (2017) An assessment of supply chain disruption mitigation strategies. Int J Prod Econ 184:210–230

Kaur H, Singh SP (2019) Sustainable procurement and logistics for disaster resilient supply chain. Ann Oper Res 283(1–2):309–354

Kilpatrick J (2020) Managing supply chain risk and disruption COVID-19. In: Deloitte. https://www2.deloitte.com/global/en/pages/risk/articles/covid-19-managing-supply-chain-risk-and-disruption.html

Lynch GS (2012) Supply chain risk management. In: Supply chain disruptions: theory and practice of managing risk, pp 319–336. 9780857297785

Mari SI, Lee YH, Memon MS (2014) Sustainable and resilient supply chain network design under disruption risks. Sustainability (Switzerland) 6(10):6666–6686

Martins VWB, Anholon R, Leal Filho W, Quelhas OLG (2021) Resilience in the supply chain management: understanding critical aspects and how digital technologies can contribute to Brazilian companies in the COVID-19 context. Modern Supply Chain Res Appl 626

Mehrjerdi YZ, Shafiee M (2021) A resilient and sustainable closed-loop supply chain using multiple sourcing and information sharing strategies. J Clean Prod 289:125141

Mensah P, Merkuryev Y (2014) Developing a resilient supply chain. Procedia Soc Behav Sci 110:309–319
Nayeri S, Sazvar Z, Heydari J (2022) A global-responsive supply chain considering sustainability and resiliency: application in the medical devices industry. Socioecon Plann Sci 82:101303
Pavlov A, Ivanov D, Pavlov D, Slinko A (2019) Optimization of network redundancy and contingency planning in sustainable and resilient supply chain resource management under conditions of structural dynamics. Ann Oper Res
Ponis ST, Koronis E (2012) Supply chain resilience: definition of concept and its formative elements. J Appl Bus Res 28(5):921–930
Ponomarov SY, Holcomb MC (2009) Understanding the concept of supply chain resilience. Int J Logist Manag 20(1):124–143
Razavian E, Alem Tabriz A, Zandieh M, Hamidizadeh MR (2021) An integrated material-financial risk-averse resilient supply chain model with a real-world application. Comput Ind Eng 161:107629
Responsible consumption and production | United Nations iLibrary (n.d.). https://www.un.org/development/desa/disabilities/envision2030-goal12.html. Accessed 27 Nov 2020
Salama MR, McGarvey RG (2021) Resilient supply chain to a global pandemic. Int J Prod Res
Shafiee M, Zare Mehrjerdi Y, Keshavarz M (2021) Integrating lean, resilient, and sustainable practices in supply chain network: mathematical modelling and the AUGMECON2 approach. Int J Syst Sci: Oper Logist
Shao XF (2013) Supply chain characteristics and disruption mitigation capability: an empirical investigation in China. Int J Log Res Appl 16(4):277–295
Shen ZM, Sun Y (2021) Strengthening supply chain resilience during COVID-19: a case study of JD.com. J Oper Manag
Singh S, Kumar R, Panchal R, Tiwari MK (2021) Impact of COVID-19 on logistics systems and disruptions in food supply chain. Int J Prod Res 59(7):1993–2008
Sodhi MS, Tang CS (2012) Managing supply chain risk. Int Ser Oper Res Manag Sci
Tang CS, Zhou S (2012) Research advances in environmentally and socially sustainable operations. Eur J Oper Res 223(3):585–594
Tukamuhabwa BR, Stevenson M, Busby J, Zorzini M (2015) Supply chain resilience: definition, review and theoretical foundations for further study. Int J Prod Res 53(18):5592–6523
Tukamuhabwa B, Stevenson M, Busby J (2017) Supply chain resilience in a developing country context: a case study on the interconnectedness of threats, strategies and outcomes. Supply Chain Manag 22(6):486–505
Udofia EE, Adejare BO, Olaore GO, Udofia EE (2021) Supply disruption in the wake of COVID-19 crisis and organisational performance: mediated by organisational productivity and customer satisfaction. J Humanit Appl Soc Sci 3(5):319–338
Youn S, Yang MG, Hong P, Park K (2013) Strategic supply chain partnership, environmental supply chain management practices, and performance outcomes: an empirical study of Korean firms. J Clean Prod 56:121–130
Zahiri B, Zhuang J, Mohammadi M (2017) Toward an integrated sustaina-ble-resilient supply chain: a pharmaceutical case study. Transp Res Part E: Logist Transp Rev. https://doi.org/10.1016/j.tre.2017.04.009
Zailani S, Jeyaraman K, Vengadasan G, Premkumar R (2012) Sustainable supply chain management (SSCM) in Malaysia: a survey. Int J Prod Econ 140(1):330–340
Zailani S, Govindan K, Iranmanesh M, Shaharudin MR (2015) Green innovation adoption in automotive supply chain: the malaysian case. J Clean Prod

GPSR Compliance
The European Union's (EU) General Product Safety Regulation (GPSR) is a set of rules that requires consumer products to be safe and our obligations to ensure this.

If you have any concerns about our products, you can contact us on

ProductSafety@springernature.com

In case Publisher is established outside the EU, the EU authorized representative is:

Springer Nature Customer Service Center GmbH
Europaplatz 3
69115 Heidelberg, Germany

www.ingramcontent.com/pod-product-compliance
Ingram Content Group UK Ltd.
Pitfield, Milton Keynes, MK11 3LW, UK
UKHW020805290926
15470UKWH00003B/958
* 9 7 8 9 8 1 9 9 2 6 2 8 2 *